AN INTRODUCTION TO

PSEUDO-DIFFERENTIAL OPERATORS

AN INTRODUCTION TO

PSEUDO-DIFFERENTIAL OPERATORS

2nd Edition

M W WONG

Department of Mathematics and Statistics
York University
Canada

World Scientific
Singapore • New Jersey • London • Hong Kong

Published by

World Scientific Publishing Co. Pte. Ltd.

P O Box 128, Farrer Road, Singapore 912805

USA office: Suite 1B, 1060 Main Street, River Edge, NJ 07661

UK office: 57 Shelton Street, Covent Garden, London WC2H 9HE

Library of Congress Cataloging-in-Publication Data
Wong, Man Wah, 1951–
    An introduction to pseudo-differential operators / M.W. Wong, --
2nd ed.
      p.   cm.
    Includes bibliographical references and index.
    ISBN 9810238134 (alk. paper)
    1. Pseudodifferential operators.   I. Title.
QA329.7.W658   1999
515'.7242--dc21
                                        98-54442
                                            CIP

British Library Cataloguing-in-Publication Data
A catalogue record for this book is available from the British Library.

This book is printed on acid-free paper.

Printed in Singapore by Uto-Print

PREFACE FOR
THE SECOND EDITION

The first edition of the book has been used as the textbook for the standard graduate course in partial differential equations at York University since its publication in 1991. The motivation for writing the second edition stems from the desire to remove several deficiencies and obscurities, and to incorporate the improvements that I can see through many years of teaching the subject to graduate students and discussions of the subject with colleagues. Notwithstanding the many changes I have in mind, I am convinced that the elementary character of the book has served and will serve well as an ideal introduction to the study of pseudo-differential operators. Thus, the basic tenet of the second edition is to retain the style and the scope of the first edition.

Notable in the second edition is the addition of two chapters to the book. Experience in teaching pseudo-differential operators reveals the fact that many graduate students are still not comfortable with the interchange of order of integration and differentiation. The new chapter added to the beginning of the book is to prove a theorem to this effect which can cope with every interchange of order of integration and differentiation encountered in the book. Another new chapter, added as the final chapter in the second edition, is to prove a theorem on the existence of weak solutions of pseudo-differential equations. The inclusion of this chapter, in my opinion, enhances the value of the book as a book on partial differential equations. Furthermore, it provides a valuable connection with the chapter on minimal and maximal operators and the chapter on global regularity.

Other new features in the second edition include a deeper study of elliptic operators and parametrices, more details on the proof of the L^p-boundedness of pseudo-differential operators, additional exercises in several chapters of the book, a slightly expanded bibliography and an index.

PREFACE

The aim of the book is to give a straightforward account of a class of pseudo-differential operators. The prerequisite for understanding the book is a course in real variables. It is hoped that the book can be used in courses in functional analysis, Fourier analysis and partial differential equations.

The first eight chapters of the book contain the basic formal calculus of pseudo-differential operators. The remaining five chapters are devoted to some topics of a more functional analytic character.

It is clear to the expert that the book takes up a single theme in a wide subject and many important topics are omitted. It is my belief that this approach is in fact a more effective introduction of pseudo-differential operators to mathematicians and graduate students beginning to learn the subject. Exercises are included in the text. They are useful to anyone who wants to understand and appreciate the book better.

The actual writing of the book was essentially carried out and completed at the University of California at Irvine while I was on sabbatical leave from York University in the academic year in 1987–88. The preliminary drafts of the book have been used in seminars and graduate courses at the University of California at Irvine and York University.

Many colleagues and students have helped me improve the contents and organization of the book. In particular, I wish to thank Professor William Margulies at the California State University at Long Beach, Professor Martin Schechter at the University of California at Irvine, Professor Tuan Vu and Mr. Zhengbin Wang at York University for their stimulating conversations and critical comments about my book. I also wish to thank Mr. Lian Pi, my Ph.D. research student at York University, who has worked out every exercise in the book.

CONTENTS

INTRODUCTION, NOTATION AND PRELIMINARIES

Let $\mathbb{R}^n$ be the usual Euclidean space given by

$$\mathbb{R}^n = \{(x_1, x_2, \ldots, x_n) : x_j\text{'s are real numbers}\}.$$

We denote points in $\mathbb{R}^n$ by x, y, ξ, η etc. Let $x = (x_1, x_2, \ldots, x_n)$ and $y = (y_1, y_2, \ldots, y_n)$ be any two points in $\mathbb{R}^n$. The inner product $x \cdot y$ of x and y is defined by

$$x \cdot y = \sum_{j=1}^{n} x_j y_j$$

and the norm $|x|$ of x is defined by

$$|x| = \left(\sum_{j=1}^{n} x_j^2 \right)^{\frac{1}{2}}.$$

On $\mathbb{R}^n$, the simplest differential operators are $\frac{\partial}{\partial x_j}, j = 1, 2, \ldots, n$. We sometimes denote $\frac{\partial}{\partial x_j}$ by ∂_j. For reasons we shall see later in the book, we usually find the operator D_j given by $D_j = -i\partial_j, i^2 = -1$, better in expressing certain formulas.

The most general linear partial differential operator of order m on $\mathbb{R}^n$ treated in this book is of the form

$$\sum_{\alpha_1+\alpha_2+\cdots+\alpha_n\leq m} a_{\alpha_1,\alpha_2,\ldots,\alpha_n}(x)D_1^{\alpha_1}D_2^{\alpha_2}\ldots D_n^{\alpha_n}, \tag{0.1}$$

where $\alpha_1,\alpha_2,\ldots,\alpha_n$ are nonnegative integers and $a_{\alpha_1,\alpha_2,\ldots,\alpha_n}(x)$ is an infinitely differentiable complex-valued function on $\mathbb{R}^n$. To simplify the expression (0.1), we let

$$\alpha = (\alpha_1,\alpha_2,\ldots,\alpha_n),$$
$$|\alpha| = \sum_{j=1}^n \alpha_j$$

and

$$D^\alpha = D_1^{\alpha_1}D_2^{\alpha_2}\ldots D_n^{\alpha_n}.$$

The α, given by an n-tuple of nonnegative integers, is called a *multi-index*. We call $|x|$ the *length* of the multi-index α. With the help of multi-indices, we can rewrite our differential operator (0.1) in the better form

$$\sum_{|\alpha|\leq m} a_\alpha(x)D^\alpha. \tag{0.2}$$

For each fixed x in $\mathbb{R}^n$, the operator (0.2) is a polynomial in $D_1,D_2,\ldots,D_n$. Therefore it is natural to denote the operator (0.2) by $P(x,D)$. If we replace D in (0.2) by a point $\xi=(\xi_1,\xi_2,\ldots,\xi_n)$ in $\mathbb{R}^n$, then we obtain a polynomial $\sum_{|\alpha|\leq m} a_\alpha(x)\xi^\alpha$ in $\mathbb{R}^n$, where $\xi^\alpha = \xi_1^{\alpha_1}\xi_2^{\alpha_2}\ldots\xi_n^{\alpha_n}$. Naturally, this polynomial is denoted by $P(x,\xi)$. We call $P(x,\xi)$ the *symbol* of the operator $P(x,D)$.

In this book, we shall study the partial differential operators (0.2) and their generalizations called pseudo-differential operators. To do this, we find it convenient to introduce in the first three chapters certain aspects of analysis pertinent to our need.

The following list of remarks, notation and formulas will be useful to us.

(i) We denote the set of all real numbers by $\mathbb{R}$ and the set of all complex numbers by $\mathbb{C}$.

(ii) All vector spaces are assumed to be over the field of complex numbers. All functions are assumed to be complex-valued unless otherwise specified.

(iii) We do not bother to distinguish a function f from its value $f(x)$ at x. In other words, we shall occasionally use the symbol $f(x)$ to denote the function f without any warning.

(iv) Although the differential operator $D^\alpha = D_1^{\alpha_1} D_2^{\alpha_2} \ldots D_n^{\alpha_n}$ is more useful to us, we still use the differential operator $\partial^\alpha = \partial_1^{\alpha_1} \partial_2^{\alpha_2} \ldots \partial_n^{\alpha_n}$ very often in the book. In case we want to emphasize the variable x (or ξ) with respect to which we differentiate, we write ∂_x^α (or ∂_ξ^α) for ∂^α and D_x^α (or D_ξ^α) for D^α.

(v) We denote the set of all infinitely differentiable functions on $\mathbb{R}^n$ by $C^\infty(\mathbb{R}^n)$.

(vi) The L^p norm of a function f in $L^p(\mathbb{R}^n)$, $1 \le p \le \infty$, is denoted by $\|f\|_p$.

Let $\alpha = (\alpha_1, \alpha_2, \ldots, \alpha_n)$ and $\beta = (\beta_1, \beta_2, \ldots, \beta_n)$ be any two multi-indices.

(vii) $\beta \le \alpha$ means that $\beta_j \le \alpha_j$ for $j = 1, 2, \ldots, n$.

(viii) $\alpha - \beta$ is the multi-index $(\alpha_1 - \beta_1, \alpha_2 - \beta_2, \ldots, \alpha_n - \beta_n)$ whenever $\beta \le \alpha$.

(ix) $\alpha! = \alpha_1! \alpha_2! \ldots \alpha_n!$.

(x) $\begin{pmatrix} \alpha \\ \beta \end{pmatrix} = \begin{pmatrix} \alpha_1 \\ \beta_1 \end{pmatrix} \begin{pmatrix} \alpha_2 \\ \beta_2 \end{pmatrix} \ldots \begin{pmatrix} \alpha_n \\ \beta_n \end{pmatrix}$ whenever $\beta \le \alpha$.

(xi) $D^\alpha(fg) = \sum\limits_{\beta \le \alpha} \begin{pmatrix} \alpha \\ \beta \end{pmatrix} (D^\beta f)(D^{\alpha-\beta} g)$.   (0.3)

The formula (0.3) is known as Leibnitz' formula. It is a special case of the following more general Leibnitz' formula.

(xii) Let $P(D) = \sum_{|\alpha| \leq m} a_\alpha D^\alpha$ be a linear partial differential operator with constant coefficients, and $P(\xi)$ its symbol. Then

$$P(D)(fg) = \sum_{|\mu| \leq m} \frac{1}{\mu!} (P^{(\mu)}(D)f)(D^\mu g) ,$$

where $P^{(\mu)}(D)$ is the linear partial differential operator with symbol $P^{(\mu)}(\xi)$ given by

$$P^{(\mu)}(\xi) = (\partial^\mu P)(\xi) , \quad \xi \in \mathbb{R}^n .$$

(xiii) $D^\alpha\left(\frac{1}{f}\right) = \sum C_{\alpha^{(1)},\alpha^{(2)},\ldots,\alpha^{(k)}} \dfrac{(\partial^{\alpha^{(1)}} f)(\partial^{\alpha^{(2)}} f)\ldots(\partial^{\alpha^{(k)}} f)}{f^{k+1}} ,$ (0.4)

where $C_{\alpha^{(1)},\alpha^{(2)},\ldots,\alpha^{(k)}}$'s are constants and the sum is taken over all possible multi-indices $\alpha^{(1)}, \alpha^{(2)}, \ldots, \alpha^{(k)}$ which form a partition of α. In (0.4), the function f is assumed to be a function in $C^\infty(\mathbb{R}^n)$. The formula (0.4) is valid at all points x in $\mathbb{R}^n$ for which $f(x) \neq 0$.

1. DIFFERENTIATION OF INTEGRALS DEPENDING ON PARAMETERS

The aim of this chapter is to prove a theorem on how to differentiate an integral depending on parameters in order to justify every interchange of integration and differentiation throughout the book. I hope analysts can find the theorem, or some variant of it, useful. Other criteria can be found in, e.g., Friedman [4, p.288], Royden [8, pp.94–95], and Wheeden and Zygmund [14, p.85].

Theorem 1.1. *Let (Y, μ) be a measure space and $f : \mathbb{R}^n \times Y \to \mathbb{C}$ be a measurable function such that*

(i) $f(x, \cdot) \in L^1(Y)$ *for all x in $\mathbb{R}^n$;*

(ii) $f(\cdot, y) \in C^\infty(\mathbb{R}^n)$ *for almost all y in Y;*

(iii) $\displaystyle \sup_{x \in \mathbb{R}^n} \int_Y |(\partial_x^\alpha f)(x, y)| d\mu < \infty$ *for all multi-indices α.*

Then the integral $\int_Y f(x, y) d\mu$, as a function of x, is in $C^\infty(\mathbb{R}^n)$ and

$$\partial^\beta \int_Y f(x, y) d\mu = \int_Y (\partial_x^\beta f)(x, y) d\mu, \quad x \in \mathbb{R}^n,$$

for all multi-indices β.

We begin with a lemma.

Lemma 1.2. *Let $f : \mathbb{R}^n \times Y \to \mathbb{C}$ be such that the hypotheses of Theorem 1.1 are satisfied. Then, for any multi-index α and $j = 1, 2, \cdots, n$, the integrals*

$$\int_Y (\partial_x^\alpha f)(x_1, \cdots, x_j, \cdots, x_n, y) d\mu$$

and

$$\int_Y |(\partial_x^\alpha f)(x_1, \cdots, x_j, \cdots, x_n, y)| d\mu,$$

as functions of x_j, are continuous on $\mathbb{R}$.

Proof: Using the mean value theorem and hypothesis (iii), we obtain

$$\int_Y \{|(\partial_x^\alpha f)(x_1, \cdots, x_j + h, \cdots, x_n, y)|-$$

$$|(\partial_x^\alpha f)(x_1, \cdots, x_j, \cdots, x_n, y)|\} d\mu$$

$$\leq M_{\alpha, j} |h|$$

for all x in $\mathbb{R}^n$, where

$$M_{\alpha, j} = \sup_{x \in \mathbb{R}^n} \int_Y |(\partial_x^\alpha \partial_{x_j} f)(x, y)| d\mu,$$

and hence the lemma.

Proof of Theorem 1.1.   In view of the proof of Lemma 1.2, the theorem is valid for the zero multi-index. Suppose that the theorem is valid for any multi-index with length l and let γ be a multi-index with length $l + 1$. If we write γ as $\beta + \varepsilon$, where β is a multi-index with length l, and ε is a multi-index with length one and the only nonzero entry in the jth position, then, by the fundamental theorem of calculus, Lemma 1.2 and the Fubini theorem,

$$\partial^\gamma \int_Y f(x,y)d\mu$$

$$= \partial_j \int_Y (\partial_x^\beta f)(x,y)d\mu$$

$$= \lim_{h \to 0} \frac{1}{h} \int_Y \{(\partial_x^\beta f)(x_1,\cdots,x_j+h,\cdots,x_n,y)-$$

$$(\partial_x^\beta f)(x_1,\cdots,x_j,\cdots,x_n,y)\}d\mu$$

$$= \lim_{h \to 0} \frac{1}{h} \int_Y \left\{ \int_{x_j}^{x_j+h} (\partial_x^\gamma f)(x_1,\cdots,s,\cdots,x_n,y)ds \right\} d\mu$$

$$= \lim_{h \to 0} \frac{1}{h} \int_{x_j}^{x_j+h} \left\{ \int_Y (\partial_x^\gamma f)(x_1,\cdots,s,\cdots,x_n,y)d\mu \right\} ds$$

$$= \int_Y (\partial_x^\gamma f)(x,y)d\mu, \quad x \in \mathbb{R}^n.$$

Thus, by induction, the proof is complete.

From the proof of Theorem 1.1, we obtain the following result.

Corollary 1.3.　*The conclusions of Theorem 1.1 remain valid if hypothesis (iii) is replaced by the hypothesis that, for every multi-index α, the integral $\int_Y (\partial_x^\alpha f)(x,y)d\mu$, as a function of x, is continuous on $\mathbb{R}^n$.*

Exercises

1.1.　Let f be a bounded function defined on the strip $Q = \{(x_1,x_2) : x_1 \in \mathbb{R},\ 0 \le x_2 \le 1\}$ in $\mathbb{R}^2$ such that for each fixed x_1 in $\mathbb{R}$, the function $f(x_1,\cdot)$ is measurable on $[0,1]$. Furthermore, suppose that $(\partial_1^k f)(x_1,x_2)$ exists for all nonnegative integers k and all $(x_1, x_2) \in Q$, and for each nonnegative integer k, there exists a positive constant C_k such that

$$\left| \left(\partial_1^k f \right) (x_1, x_2) \right| \leq C_k, \quad (x_1, x_2) \in Q .$$

Prove that

$$\left(\frac{d}{dx_1} \right)^k \int_0^1 f(x_1, x_2) dx_1 = \int_0^1 \left(\partial_1^k f \right) (x_1, x_2) dx_1$$

for all $x_2 \in \mathbb{R}$.

1.2.   Let $f \in L^1(\mathbb{R}^n)$ and let $g \in C^\infty(\mathbb{R}^n)$ be such that $\partial^\alpha g \in L^\infty(\mathbb{R}^n)$ for all multi-indices α. Prove that the function h on $\mathbb{R}^n$ defined by

$$h(x) = \int_{\mathbb{R}^n} f(y) g(x - y) dy , \quad x \in \mathbb{R}^n ,$$

is in $C^\infty(\mathbb{R}^n)$ and

$$\left(\partial^\alpha h \right) (x) = \int_{\mathbb{R}^n} f(y) \left(\partial^\alpha g \right) (x - y) dy , \quad x \in \mathbb{R}^n .$$

1.3.   Let f be the function on $\mathbb{R}$ defined by

$$f(x) = \sum_{n=-\infty}^{\infty} c_n e^{inx} , \quad x \in \mathbb{R} ,$$

where $\cdots, c_{-2}, c_{-1}, c_0, c_1, c_2, \cdots$ are constants such that $\sum_{n=-\infty}^{\infty} |c_n| < \infty$. Prove that, if $\sum_{n=-\infty}^{\infty} |c_n| n^k < \infty$ for all nonnegative integers k, then $f \in C^\infty(\mathbb{R})$ and, for $k = 0, 1, 2, \cdots$, we have

$$f^{(k)}(x) = \sum_{n=-\infty}^{\infty} c_n n^k e^{inx} , \quad x \in \mathbb{R} .$$

2. THE CONVOLUTION

In this chapter, we introduce two important subsets of $C^\infty(\mathbb{R}^n)$, usually denoted by $C_0^\infty(\mathbb{R}^n)$ and $\mathcal{S}$. The aim of this chapter is to prove that they are dense in $L^p(\mathbb{R}^n), 1 \leq p < \infty$. To this end, we need the notion of convolution.

Theorem 2.1. **(Young's Inequality)** *Let $f \in L^1(\mathbb{R}^n)$ and $g \in L^p(\mathbb{R}^n), 1 \leq p \leq \infty$. Then the integral*

$$\int_{\mathbb{R}^n} f(x-y)g(y)dy$$

exists for almost every $x \in \mathbb{R}^n$. If the value of the integral is denoted by $(f * g)(x)$, then $f * g \in L^p(\mathbb{R}^n)$ and

$$\|f * g\|_p \leq \|f\|_1 \|g\|_p .$$

Remark 2.2. We usually call $f * g$ *the convolution of f and g.*

Proof of Theorem 2.1. For $p = 1$, let

$$h(x) = \int_{\mathbb{R}^n} |f(x-y)| \, |g(y)| dy .$$

Then, by Fubini's theorem,

$$\int_{\mathbb{R}^n} h(x)dx = \int_{\mathbb{R}^n} \left(\int_{\mathbb{R}^n} |f(x-y)| \, |g(y)|dy \right) dx$$

$$= \int_{\mathbb{R}^n} |g(y)|dy \int_{\mathbb{R}^n} |f(x-y)|dx$$

$$= \|g\|_1 \|f\|_1 .$$

Hence $h(x) < \infty$ for almost every $x \in \mathbb{R}^n$. Using Fubini's theorem again, we have

$$\|f * g\|_1 = \int_{\mathbb{R}^n} \left| \int_{\mathbb{R}^n} f(x-y)g(y)dy \right| dx$$

$$\leq \int_{\mathbb{R}^n} \left(\int_{\mathbb{R}^n} |f(x-y)| \, |g(y)|dy \right) dx$$

$$= \int_{\mathbb{R}^n} |g(y)|dy \int_{\mathbb{R}^n} |f(x-y)|dx$$

$$= \|g\|_1 \|f\|_1 .$$

This proves the theorem for $p = 1$. For $1 < p < \infty$, let

$$h_p(x) = \int_{\mathbb{R}^n} |f(x-y)| \, |g(y)|^p dy .$$

Then, as in the previous case, $h_p(x) < \infty$ for almost every $x \in \mathbb{R}^n$. Let p' be the conjugate index of p. Then, by Hölder's inequality,

$$\int_{\mathbb{R}^n} |f(x-y)| \, |g(y)|dy$$

$$= \int_{\mathbb{R}^n} |f(x-y)|^{\frac{1}{p'}} |f(x-y)|^{\frac{1}{p}} |g(y)|dy$$

$$\leq \left\{ \int_{\mathbb{R}^n} |f(x-y)|dy \right\}^{\frac{1}{p'}} \left\{ \int_{\mathbb{R}^n} |f(x-y)| \, |g(y)|^p dy \right\}^{\frac{1}{p}}$$

$$= \|f\|_1^{\frac{1}{p'}} \{h_p(x)\}^{\frac{1}{p}} . \tag{2.1}$$

Hence $\int_{\mathbb{R}^n} f(x-y)g(y)dy$ exists for almost every $x \in \mathbb{R}^n$. Moreover, by (2.1), we have

$$
\begin{aligned}
\|f * g\|_p &= \left\{ \int_{\mathbb{R}^n} \left| \int_{\mathbb{R}^n} f(x-y)g(y)dy \right|^p dx \right\}^{\frac{1}{p}} \\
&\leq \left\{ \int_{\mathbb{R}^n} \left(\int_{\mathbb{R}^n} |f(x-y)|\, |g(y)|\, dy \right)^p dx \right\}^{\frac{1}{p}} \\
&\leq \|f\|_1^{\frac{1}{p'}} \left\{ \int_{\mathbb{R}^n} h_p(x)dx \right\}^{\frac{1}{p}} \\
&\leq \|f\|_1^{\frac{1}{p'}} \left\{ \|f\|_1\, \|G\|_1 \right\}^{\frac{1}{p}},
\end{aligned}
$$

where $G(x) = |g(x)|^p$ for all $x \in \mathbb{R}^n$. Hence

$$
\|f * g\|_p \leq \|f\|_1\, \|g\|_p .
$$

To prove the theorem for $p = \infty$, note that

$$
\int_{\mathbb{R}^n} |f(x-y)|\, |g(y)|\, dy \leq \|g\|_\infty \int_{\mathbb{R}^n} |f(x-y)|dy = \|g\|_\infty \|f\|_1 .
$$
(2.2)

Hence the integral $\int_{\mathbb{R}^n} f(x-y)g(y)dy$ exists for every $x \in \mathbb{R}^n$. Moreover, by (2.2), we have

$$
\|f * g\|_\infty \leq \|g\|_\infty \|f\|_1 .
$$

Proposition 2.3. (*L^p-Continuity of Translations*) *Let $f \in L^p(\mathbb{R}^n)$, $1 \leq p < \infty$. Then*

$$
\lim_{|x| \to 0} \|f_x - f\|_p = 0 ,
$$

where f_x is the function defined by
$$
f_x(y) = f(x+y) , \quad y \in \mathbb{R}^n .
$$

Before proving Proposition 2.3, let us define $C_0(\mathbb{R}^n)$ to be the set of all continuous functions on $\mathbb{R}^n$ with compact supports. The support of a continuous function h is defined to be the closure in $\mathbb{R}^n$ of the set

$$\{x \in \mathbb{R}^n \ : \ h(x) \neq 0\}$$

and is denoted by supp(h). We give one property of the set $C_0(\mathbb{R}^n)$ in the following proposition:

Proposition 2.4. $C_0(\mathbb{R}^n)$ *is dense in* $L^p(\mathbb{R}^n)$ *for* $1 \leq p < \infty$.

Proposition 2.4 is a measure theoretic result which I ask you to believe. Otherwise, see page 69 in Rudin [9] for a proof.

Proof of Proposition 2.3. Let $\delta > 0$ and $f \in L^p(\mathbb{R}^n)$. Then, by Proposition 2.4, there is a function g in $C_0(\mathbb{R}^n)$ such that

$$\|f - g\|_p < \frac{\delta}{3} . \tag{2.3}$$

Now, using the triangle inequality and (2.3),

$$\|f_x - f\|_p \leq \|f_x - g_x\|_p + \|g_x - g\|_p + \|g - f\|_p < \frac{\delta}{3} + \frac{\delta}{3} + \frac{\delta}{3} = \delta$$

if $|x|$ is small enough. This completes the proof.

Theorem 2.5. *Let* $\varphi \in L^1(\mathbb{R}^n)$ *be such that* $\int_{\mathbb{R}^n} \varphi(x)dx = a$. *For* $\varepsilon > 0$, *define the function* φ_ε *by*

$$\varphi_\varepsilon(x) = \varepsilon^{-n} \varphi \left(\frac{x}{\varepsilon} \right) , \quad x \in \mathbb{R}^n .$$

Then, for any function $f \in L^p(\mathbb{R}^n), 1 \leq p < \infty$, *we have* $f * \varphi_\varepsilon \to af$ *in* $L^p(\mathbb{R}^n)$ *as* $\varepsilon \to 0$.

Proof: Since

$$\int_{\mathbb{R}^n} \varphi_\varepsilon(x)dx = a$$

for all $\varepsilon > 0$, it follows from Minkowski's inequality in integral form that we have

$$\|f * \varphi_\varepsilon - af\|_p = \left(\int_{\mathbb{R}^n} |(f * \varphi_\varepsilon)(x) - af(x)|^p \, dx \right)^{\frac{1}{p}}$$

$$= \left(\int_{\mathbb{R}^n} \left| \int_{\mathbb{R}^n} \{f(x-y) - f(x)\} \varphi_\varepsilon(y) dy \right|^p dx \right)^{\frac{1}{p}}$$

$$= \left(\int_{\mathbb{R}^n} \left| \int_{\mathbb{R}^n} \{f(x-\varepsilon y) - f(x)\} \varphi(y) dy \right|^p dx \right)^{\frac{1}{p}}$$

$$\leq \int_{\mathbb{R}^n} \left\{ \int_{\mathbb{R}^n} |f(x-\varepsilon y) - f(x)|^p |\varphi(y)|^p dx \right\}^{\frac{1}{p}} dy$$

$$= \int_{\mathbb{R}^n} |\varphi(y)| \left\{ \int_{\mathbb{R}^n} |f(x-\varepsilon y) - f(x)|^p \, dx \right\}^{\frac{1}{p}} dy$$

$$= \int_{\mathbb{R}^n} |\varphi(y)| \, \|f_{-\varepsilon y} - f\|_p dy . \qquad (2.4)$$

By Proposition 2.3, $\|f_{-\varepsilon y} - f\|_p \to 0$ as $\varepsilon \to 0$. Also, by the triangle inequality, $\|f_{-\varepsilon y} - f\|_p \leq 2\|f\|_p$. Hence an application of the Lebsegue dominated convergence theorem to the last integral in (2.4) implies that

$$\|f * \varphi_\varepsilon - af\|_p \to 0 \quad \text{as} \quad \varepsilon \to 0 .$$

We introduce two important function spaces. It is customary to denote by $C_0^\infty(\mathbb{R}^n)$ the set of all infinitely differentiable functions on $\mathbb{R}^n$ with compact supports and by $\mathcal{S}$ the set of all infinitely differentiable functions φ on $\mathbb{R}^n$ such that for all multi-indices α and β,

$$\sup_{x \in \mathbb{R}^n} \left| x^\alpha (D^\beta \varphi)(x) \right| < \infty .$$

The space $\mathcal{S}$ is usually called the *Schwartz space* in deference to Laurent Schwartz. Obviously, $C_0^\infty(\mathbb{R}^n)$ is included in $\mathcal{S}$. That the inclusion is proper can be seen easily by noting that the function

$e^{-|x|^2}$ is in S but not in $C_0^\infty(\mathbb{R}^n)$. We want to prove that $C_0^\infty(\mathbb{R}^n)$ is dense in $L^p(\mathbb{R}^n)$ for $1 \le p < \infty$. To this end, we need two preliminary results.

Proposition 2.6.   *Let $\varphi \in S$ and $f \in L^p(\mathbb{R}^n), 1 \le p \le \infty$. Then $f * \varphi \in C^\infty(\mathbb{R}^n)$ and*

$$\partial^\alpha(f * \varphi) = f * \partial^\alpha \varphi$$

for every multi-index α.

Proof: Let $\varphi \in S$. Then, for every multi-index α, $\partial^\alpha \varphi \in L^{p'}(\mathbb{R}^n)$, where p' is the conjugate index of p. Hence, by Hölder's inequality, $(f * \partial^\alpha \varphi)(x)$ exists for every $x \in \mathbb{R}^n$. Therefore differentiation and integration can be interchanged.

Proposition 2.7.   *Let f and g be continuous functions on $\mathbb{R}^n$ with compact supports. Then the convolution $f * g$ also has compact support. In fact,*

$$\operatorname{supp}(f * g) \subseteq \operatorname{supp}(f) + \operatorname{supp}(g) .$$

Remark 2.8.   The vector sum $A + B$ of two sets A and B is defined by

$$A + B = \{x + y : x \in A \quad \text{and} \quad y \in B\} .$$

Proof of Proposition 2.7.   Since

$$(f * g)(x) = \int_{\mathbb{R}^n} f(x - y)g(y)dy ,$$

it follows that if $(f * g)(x) \neq 0$, then there exists a $y \in \operatorname{supp}(g)$ such that $x - y \in \operatorname{supp}(f)$. Hence, by Remark 2.8, $x \in \operatorname{supp}(f) + \operatorname{supp}(g)$.

Theorem 2.9. $C_0^\infty(\mathbb{R}^n)$ *is dense in* $L^p(\mathbb{R}^n)$ *for* $1 \le p < \infty$.

Proof: Let $\varphi \in C_0^\infty(\mathbb{R}^n)$ be such that $\int_{\mathbb{R}^n} \varphi(x)dx = 1$. Such a function exists by Exercise 2.2. For $\varepsilon > 0$, define φ_ε by

$$\varphi_\varepsilon(x) = \varepsilon^{-n}\varphi\left(\frac{x}{\varepsilon}\right), \quad x \in \mathbb{R}^n .$$

Then, for all functions $g \in C_0(\mathbb{R}^n)$, we have, by Propositions 2.6 and 2.7, $g * \varphi_\varepsilon \in C_0^\infty(\mathbb{R}^n)$. Also, by Theorem 2.5,

$$g * \varphi_\varepsilon \to g \tag{2.5}$$

in $L^p(\mathbb{R}^n)$ as $\varepsilon \to 0$. Let $\delta > 0$ and $f \in L^p(\mathbb{R}^n)$. Then, by Proposition 2.4, there is a function $g \in C_0(\mathbb{R}^n)$ such that

$$\|f - g\|_p < \frac{\delta}{2} . \tag{2.6}$$

By (2.5), we can find a function $\psi \in C_0^\infty(\mathbb{R}^n)$ such that

$$\|g - \psi\|_p < \frac{\delta}{2} . \tag{2.7}$$

Hence, by the triangle inequality, (2.6) and (2.7), we have

$$\|f - \psi\|_p \le \|f - g\|_p + \|g - \psi\|_p < \frac{\delta}{2} + \frac{\delta}{2} = \delta .$$

This proves that $C_0^\infty(\mathbb{R}^n)$ is dense in $L^p(\mathbb{R}^n)$ for $1 \le p < \infty$.

Remark 2.10. An immediate consequence of Exercise 2.3 and Theorem 2.9 is that the Schwartz space $\mathcal{S}$ is also dense in $L^p(\mathbb{R}^n)$ for $1 \le p < \infty$.

Exercises

2.1.   Let φ and φ_ε be the functions given in the hypotheses of Theorem 2.5. Let f be a bounded function on $\mathbb{R}^n$ which is continuous on an open subset V of $\mathbb{R}^n$. Prove that $f * \varphi_\varepsilon \to af$ uniformly on every compact subset of V as $\varepsilon \to 0$.

2.2.   Let φ be the function on $\mathbb{R}^n$ defined by

$$\varphi(x) = \begin{cases} \exp\left(-\frac{1}{1-|x|^2}\right), & |x| < 1, \\ 0, & |x| \geq 1. \end{cases}$$

Prove that $\varphi \in C_0^\infty(\mathbb{R}^n)$.

2.3.   Prove that every function in $\mathcal{S}$ is in $L^p(\mathbb{R}^n)$, $1 \leq p \leq \infty$.

2.4.   Is $\mathcal{S}$ dense in $L^\infty(\mathbb{R}^n)$? Explain your answer.

3.   THE FOURIER TRANSFORM

The Fourier transform will be used in Chapter 5 to define pseudo-differential operators. Two important results in the theory of Fourier transforms are the Fourier inversion formula for Schwartz functions in S and the Plancherel formula for functions in $L^2(\mathbb{R}^n)$. They are very useful for the study of pseudo-differential operators.

Let $f \in L^1(\mathbb{R}^n)$. We define $\hat{f}$ by

$$\hat{f}(\xi) = (2\pi)^{-n/2} \int_{\mathbb{R}^n} e^{-ix\cdot\xi} f(x)dx \ , \quad \xi \in \mathbb{R}^n \ .$$

The function $\hat{f}$ is called the *Fourier transform* of f and is sometimes denoted by $\mathcal{F}f$.

Proposition 3.1. *Let f and g be in $L^1(\mathbb{R}^n)$. Then*

$$(f * g)\hat{} = (2\pi)^{n/2} \hat{f}\hat{g} \ .$$

Proof: By Theorem 2.1, $f * g \in L^1(\mathbb{R}^n)$. Then, using the definition of the Fourier transform and Fubini's theorem, we have

$$(2\pi)^{-n/2}(f * g)\hat{\ }(\xi)$$

$$= (2\pi)^{-n} \int_{\mathbb{R}^n} e^{-ix\cdot\xi}(f * g)(x)dx$$

$$= (2\pi)^{-n} \int_{\mathbb{R}^n} e^{-ix\cdot\xi}\left(\int_{\mathbb{R}^n} f(x-y)g(y)dy\right)dx$$

$$= (2\pi)^{-n} \int_{\mathbb{R}^n}\int_{\mathbb{R}^n} e^{-i(x-y)\cdot\xi} f(x-y)e^{-iy\cdot\xi}g(y)dydx$$

$$= (2\pi)^{-n} \int_{\mathbb{R}^n} e^{-iy\cdot\xi}g(y)\left(\int_{\mathbb{R}^n} e^{-i(x-y)\cdot\xi} f(x-y)dx\right)dy$$

$$= (2\pi)^{-n/2}\left(\int_{\mathbb{R}^n} e^{-iy\cdot\xi}g(y)dy\right)\hat{f}(\xi)$$

$$= \hat{g}(\xi)\hat{f}(\xi) .$$

Proposition 3.2.   *Let $\varphi \in \mathcal{S}$. Then*
   (i) $(D^\alpha\varphi)\hat{\ }(\xi) = \xi^\alpha\hat{\varphi}(\xi)$ *for every multi-index α,*
   (ii) $(D^\beta\hat{\varphi})(\xi) = ((-x)^\beta\varphi)\hat{\ }(\xi)$ *for every multi-index β,*
   (iii) $\hat{\varphi} \in \mathcal{S}$.

Proof: Integrating by parts, we get

$$(D^\alpha\varphi)\hat{\ }(\xi) = (2\pi)^{-n/2} \int_{\mathbb{R}^n} e^{-ix\cdot\xi}(D^\alpha\varphi)(x)dx$$

$$= (2\pi)^{-n/2} \int_{\mathbb{R}^n} \xi^\alpha e^{-ix\cdot\xi}\varphi(x)dx$$

$$= \xi^\alpha\hat{\varphi}(\xi) .$$

This proves part **(i)**. For part **(ii)**, we have

$$\left(D^\beta\hat{\varphi}\right)(\xi) = (2\pi)^{-n/2}D^\beta\left(\int_{\mathbb{R}^n} e^{-ix\cdot\xi}\varphi(x)dx\right)$$

$$= (2\pi)^{-n/2} \int_{\mathbb{R}^n} (-x)^\beta e^{-ix\cdot\xi}\varphi(x)dx$$

$$= \left((-x)^\beta\varphi\right)\hat{\ }(\xi) .$$

The interchange of the order of differentiation and integration is valid because $(-x)^\beta \varphi \in \mathcal{S}$. To prove part **(iii)**, let α and β be any two multi-indices. Then, by parts **(i)** and **(ii)**,

$$\left| \xi^\alpha \left(D^\beta \hat{\varphi} \right) (\xi) \right| = \left| \xi^\alpha \widehat{\left((-x)^\beta \varphi \right)} (\xi) \right|$$

$$= \left| \left\{ D^\alpha \left((-x)^\beta \varphi \right) \right\}^\wedge (\xi) \right| .$$

Since $D^\alpha ((-x)^\beta \varphi)$ is in $\mathcal{S}$, hence in $L^1(\mathbb{R}^n)$, it follows that

$$\sup_{\xi \in \mathbb{R}^n} \left| \xi^\alpha \left(D^\beta \hat{\varphi} \right) (\xi) \right| = \sup_{\xi \in \mathbb{R}^n} \left| \left\{ D^\alpha \left((-x)^\beta \varphi \right) \right\}^\wedge (\xi) \right|$$

$$\leq (2\pi)^{-n/2} \| D^\alpha \left((-x)^\beta \varphi \right) \|_1$$

$$< \infty .$$

Proposition 3.3. **(The Riemann-Lebesgue Lemma)** *Let $f \in L^1(\mathbb{R}^n)$. Then*

   (i) *$\hat{f}$ is continuous on $\mathbb{R}^n$,*

  (ii) *$\lim_{|\xi| \to \infty} \hat{f}(\xi) = 0$,*

 (iii) *$f_j \to f$ in $L^1(\mathbb{R}^n) \Rightarrow \hat{f}_j \to \hat{f}$ uniformly on $\mathbb{R}^n$.*

Proof: Let $f_j \to f$ in $L^1(\mathbb{R}^n)$. Then

$$|\hat{f}_j(\xi) - \hat{f}(\xi)| \leq (2\pi)^{-n/2} \| f_j - f \|_1 .$$

Hence $\hat{f}_j \to \hat{f}$ uniformly on $\mathbb{R}^n$. This proves part **(iii)**. To prove parts **(i)** and **(ii)**, let $\varphi \in \mathcal{S}$. Then, by part **(iii)** of Proposition 3.2, $\hat{\varphi} \in \mathcal{S}$. Hence parts **(i)** and **(ii)** are satisfied for functions in $\mathcal{S}$. Let $f \in L^1(\mathbb{R}^n)$. Since $\mathcal{S}$ is dense in $L^1(\mathbb{R}^n)$, it follows that there is a sequence $\{\varphi_j\}$ of functions in $\mathcal{S}$ such that $\varphi_j \to f$ in $L^1(\mathbb{R}^n)$. By part **(iii)** which we have proved, $\hat{\varphi}_j \to \hat{f}$ uniformly on $\mathbb{R}^n$. This proves parts **(i)** and **(ii)**.

Let f be a measurable function defined on $\mathbb{R}^n$. For any fixed $y \in \mathbb{R}^n$, we define functions $T_y f$ and $M_y f$ by

$$(T_y f)(x) = f(x + y) , \quad x \in \mathbb{R}^n , \tag{3.1}$$

and

$$(M_y f)(x) = e^{ix \cdot y} f(x) , \quad x \in \mathbb{R}^n . \tag{3.2}$$

Let a be a nonzero real number. Then we define the function $D_a f$ by

$$(D_a f)(x) = f(ax) , \quad x \in \mathbb{R}^n . \tag{3.3}$$

Proposition 3.4.   *Let $f \in L^1(\mathbb{R}^n)$. Then the functions $T_y f, M_y f$ and $D_a f$ defined by (3.1), (3.2) and (3.3) respectively are in $L^1(\mathbb{R}^n)$. Moreover,*

(i) $(T_y f)\hat{\,}(\xi) = (M_y \hat{f})(\xi) , \xi \in \mathbb{R}^n$,

(ii) $(M_y f)\hat{\,}(\xi) = (T_{-y} \hat{f}) , \xi \in \mathbb{R}^n$,

(iii) $(D_a f)\hat{\,}(\xi) = |a|^{-n} (D_{\frac{1}{a}} \hat{f})(\xi) , \xi \in \mathbb{R}^n$.

Proof: Obviously, $T_y f, M_y f$ and $D_a f$ are in $L^1(\mathbb{R}^n)$. By a simple change of variable, we have

$$
\begin{aligned}
(T_y f)\hat{\,}(\xi) &= (2\pi)^{-n/2} \int_{\mathbb{R}^n} e^{-ix \cdot \xi} (T_y f)(x) dx \\
&= (2\pi)^{-n/2} \int_{\mathbb{R}^n} e^{-ix \cdot \xi} f(x + y) dx \\
&= (2\pi)^{-n/2} \int_{\mathbb{R}^n} e^{-i(x-y) \cdot \xi} f(x) dx \\
&= e^{iy \cdot \xi} (2\pi)^{-n/2} \int_{\mathbb{R}^n} e^{-ix \cdot \xi} f(x) dx \\
&= e^{iy \cdot \xi} \hat{f}(\xi) \\
&= (M_y \hat{f})(\xi) .
\end{aligned}
$$

Also,

$$(M_y f)\hat{}(\xi) = (2\pi)^{-n/2} \int_{\mathbb{R}^n} e^{-ix\cdot\xi} (M_y f)(x) dx$$

$$= (2\pi)^{-n/2} \int_{\mathbb{R}^n} e^{-ix\cdot\xi} e^{iy\cdot x} f(x) dx$$

$$= \hat{f}(\xi - y)$$

$$= (T_{-y}\hat{f})(\xi) .$$

Finally, by another change of variable, we have

$$(D_a f)\hat{}(\xi) = (2\pi)^{-n/2} \int_{\mathbb{R}^n} e^{-ix\cdot\xi} (D_a f)(x) dx$$

$$= (2\pi)^{-n/2} \int_{\mathbb{R}^n} e^{-ix\cdot\xi} f(ax) dx$$

$$= (2\pi)^{-n/2} \int_{\mathbb{R}^n} e^{-i(\frac{x}{a})\cdot\xi} f(x) |a|^{-n} dx$$

$$= |a|^{-n} \hat{f}\left(\frac{\xi}{a}\right)$$

$$= |a|^{-n} \left(D_{\frac{1}{a}}\hat{f}\right)(\xi) .$$

Proposition 3.5. *Let* $\varphi(x) = e^{-\frac{|x|^2}{2}}$. *Then* $\hat{\varphi}(\xi) = e^{-\frac{|\xi|^2}{2}}$.

Proof: We first compute

$$(2\pi)^{-n/2} \int_{\mathbb{R}^n} e^{-ix\cdot\xi - |x|^2} dx .$$

Note that

$$(2\pi)^{-n/2} \int_{\mathbb{R}^n} e^{-ix\cdot\xi - |x|^2} dx$$

$$= \prod_{j=1}^{n} (2\pi)^{-1/2} \int_{-\infty}^{\infty} e^{-ix_j\xi_j - x_j^2} dx_j . \tag{3.4}$$

Hence it is sufficient to compute

$$(2\pi)^{-1/2} \int_{-\infty}^{\infty} e^{-it\zeta - t^2} \, dt \, , \quad \zeta \in (-\infty, \infty) \, .$$

But

$$
\begin{aligned}
\int_{-\infty}^{\infty} e^{-it\zeta - t^2} \, dt &= \int_{-\infty}^{\infty} e^{-\left(t^2 + it\zeta\right)} \, dt \\
&= e^{-\frac{\zeta^2}{4}} \int_{-\infty}^{\infty} e^{-\left(t^2 + it\zeta - \frac{\zeta^2}{4}\right)} \, dt \\
&= e^{-\frac{\zeta^2}{4}} \int_{-\infty}^{\infty} e^{-\left(t + i\frac{\zeta}{2}\right)^2} \, dt \\
&= e^{-\frac{\zeta^2}{4}} \int_{L} e^{-z^2} \, dz \, ,
\end{aligned}
\tag{3.5}
$$

where L is the contour $\mathrm{Im}\, z = \frac{\zeta}{2}$ in the complex z-plane. Using Cauchy's integral theorem and the fact that the integrand goes to zero very fast as $|z| \to \infty$, we have

$$\int_{L} e^{-z^2} \, dz = \int_{-\infty}^{\infty} e^{-t^2} \, dt = \sqrt{\pi} \, . \tag{3.6}$$

Hence, by (3.5) and (3.6),

$$(2\pi)^{-1/2} \int_{-\infty}^{\infty} e^{-it\zeta - t^2} \, dt = 2^{-1/2} e^{-\frac{\zeta^2}{4}} \, . \tag{3.7}$$

By (3.4) and (3.7), we get

$$(2\pi)^{-n/2} \int_{\mathbb{R}^n} e^{-ix \cdot \xi - |x|^2} \, dx = 2^{-\frac{n}{2}} e^{-\frac{|\xi|^2}{4}} \, . \tag{3.8}$$

Now, note that

$$(2\pi)^{-n/2} \int_{\mathbb{R}^n} e^{-ix \cdot \xi - \frac{|x|^2}{2}} \, dx = \left(D_{\frac{1}{\sqrt{2}}} \psi \right)^{\widehat{\ }} (\xi) \, ,$$

where $\psi(x) = e^{-|x|^2}$. Therefore, by (3.8) and part (iii) of Proposition 3.4, we get

$$\hat{\varphi}(\xi) = (2\pi)^{-n/2} \int_{\mathbb{R}^n} e^{-ix\cdot\xi - \frac{|x|^2}{2}} dx = e^{-\frac{|\xi|^2}{2}} .$$

Proposition 3.6. (The Adjoint Formula) *Let f and g be functions in $L^1(\mathbb{R}^n)$. Then*

$$\int_{\mathbb{R}^n} \hat{f}(x)g(x)dx = \int_{\mathbb{R}^n} f(x)\hat{g}(x)dx . \qquad (3.9)$$

Proof: By Proposition 3.3, the Fourier transform of a function in $L^1(\mathbb{R}^n)$ is bounded on $\mathbb{R}^n$. Hence the integrals in (3.9) exist. Moreover,

$$\int_{\mathbb{R}^n} \hat{f}(x)g(x)dx = (2\pi)^{-n/2} \int_{\mathbb{R}^n} \left(\int_{\mathbb{R}^n} e^{-ix\cdot y} f(y)dy \right) g(x)dx$$

$$= (2\pi)^{-n/2} \int_{\mathbb{R}^n} f(y) \left(\int_{\mathbb{R}^n} e^{-ix\cdot y} g(x)dx \right) dy$$

$$= \int_{\mathbb{R}^n} f(y)\hat{g}(y)dy .$$

The interchange of the order of integration can obviously be justified by Fubini's theorem.

We are now prepared to prove the first important result in the theory of the Fourier transform.

Theorem 3.7. (The Fourier Inversion Formula) $(\hat{f})^{\vee} = f$ *for all functions $f \in S$. Here, the operation $\vee$ is defined by*

$$\check{g}(x) = (2\pi)^{-n/2} \int_{\mathbb{R}^n} e^{ix\cdot\xi} g(\xi)d\xi , \quad g \in S .$$

Remark 3.8. The function $\check{g}$ is usually called the *inverse Fourier transform* of g.

Proof of Theorem 3.7.   We have

$$(\hat{f})^{\vee}(x) = (2\pi)^{-n/2} \int_{\mathbb{R}^n} e^{ix\cdot\xi} \hat{f}(\xi) d\xi \ .$$

Let $\varepsilon > 0$. Define

$$I_{\varepsilon}(x) = (2\pi)^{-n/2} \int_{\mathbb{R}^n} e^{ix\cdot\xi - \frac{\varepsilon^2 |\xi|^2}{2}} \hat{f}(\xi) d\xi \ . \tag{3.10}$$

Let

$$g(\xi) = e^{ix\cdot\xi - \frac{\varepsilon^2 |\xi|^2}{2}} = \left(M_x D_\varepsilon \varphi\right)(\xi) \ , \tag{3.11}$$

where

$$\varphi(\xi) = e^{\frac{-|\xi|^2}{2}} \ . \tag{3.12}$$

Then, by Propositions 3.4 and 3.5,

$$\hat{g}(\eta) = \left(T_{-x}\varepsilon^{-n} D_{\frac{1}{\varepsilon}} \hat{\varphi}\right)(\eta)$$
$$= \varepsilon^{-n} e^{\frac{-|\eta - x|^2}{2\varepsilon^2}} \ . \tag{3.13}$$

Hence, by (3.10), (3.11), (3.13) and Proposition 3.6,

$$I_{\varepsilon}(x) = (2\pi)^{-n/2} \int_{\mathbb{R}^n} g(\xi) \hat{f}(\xi) d\xi$$
$$= (2\pi)^{-n/2} \int_{\mathbb{R}^n} \hat{g}(\eta) f(\eta) d\eta$$
$$= \varepsilon^{-n} (2\pi)^{-n/2} \int_{\mathbb{R}^n} e^{\frac{-|\eta - x|^2}{2\varepsilon^2}} f(\eta) d\eta$$
$$= (2\pi)^{-n/2} \left(f * \varphi_\varepsilon\right)(x) \ , \tag{3.14}$$

where $\varphi_\varepsilon(x) = \varepsilon^{-n} \varphi\left(\dfrac{x}{\varepsilon}\right)$. Since $f \in \mathcal{S}$, it follows that f is in $L^p(\mathbb{R}^n), 1 \le p < \infty$. Therefore, by (3.12), (3.14) and Theorem 2.5,

$$I_{\varepsilon} \to (2\pi)^{-n/2} \left(\int_{\mathbb{R}^n} e^{\frac{-|x|^2}{2}} dx\right) f = f$$

in $L^p(\mathbb{R}^n)$ as $\varepsilon \to 0$. Hence there exists a sequence $\{\varepsilon_n\}$ of positive real numbers such that $I_{\varepsilon_n}(x) \to f(x)$ for almost every $x \in \mathbb{R}^n$ as $\varepsilon_n \to 0$. By (3.10) and Lebesgue's dominated convergence theorem,

$$I_\varepsilon(x) \to (2\pi)^{-n/2} \int_{\mathbb{R}^n} e^{ix\cdot\xi} \hat{f}(\xi) d\xi$$

for every $x \in \mathbb{R}^n$ as $\varepsilon \to 0$. Hence

$$(2\pi)^{-n/2} \int_{\mathbb{R}^n} e^{ix\cdot\xi} \hat{f}(\xi) d\xi = f(x)$$

for every $x \in \mathbb{R}^n$. This proves the theorem.

Remark 3.9. An immediate consequence of the Fourier inversion formula is that the Fourier transformation $f \to \hat{f}$ is a one-to-one mapping of $\mathcal{S}$ onto $\mathcal{S}$. If we define $\tilde{f}$ by

$$\tilde{f}(x) = f(-x) , \quad x \in \mathbb{R}^n ,$$

then the Fourier inversion formula is equivalent to the formula

$$\hat{\hat{f}} = \tilde{f} , \quad f \in \mathcal{S} .$$

The next important result is the Plancherel Theorem. Before we come to it, let us prove the following proposition by using the fact that $|x^\alpha| \le |x|^{|\alpha|}$ for all $x \in \mathbb{R}^n$ and multi-indices α. The proof of this simple inequality is left as an exercise. See Exercise 3.5.

Proposition 3.10. *Let f and g be functions in $\mathcal{S}$. Then the convolution $f * g$ is also in $\mathcal{S}$.*

Proof: By Proposition 2.6, $f * g \in C^\infty(\mathbb{R}^n)$. Let α and β be multi-indices. Then, by Exercise 3.5, we have

$$\left| x^\alpha \left\{ \partial^\beta (f * g) \right\} (x) \right|$$

$$\le \int_{\mathbb{R}^n} |x - y + y|^{|\alpha|} \left| (\partial^\beta f)(x - y) \right| |g(y)| dy$$

$$\le 2^{|\alpha|} \int_{\mathbb{R}^n} |x - y|^{|\alpha|} \left| (\partial^\beta f)(x - y) \right| |g(y)| dy$$

$$+ 2^{|\alpha|} \int_{\mathbb{R}^n} |y|^{|\alpha|} \left| (\partial^\beta f)(x - y) \right| |g(y)| dy$$

for all $x \in \mathbb{R}^n$. Since f and g are in $\mathcal{S}$, it follows that

$$\sup_{x \in \mathbb{R}^n} |x^\alpha \{\partial^\beta (f * g)\}(x)| < \infty$$

and this completes the proof.

Theorem 3.11. (The Plancherel Theorem)   *The mapping* $f \to \hat{f}$ *defined on* $\mathcal{S}$ *can be extended uniquely to a unitary operator on* $L^2(\mathbb{R}^n)$.

Proof: Using the fact that $\mathcal{S}$ is dense in $L^2(\mathbb{R}^n)$ and the Fourier inversion formula, it is sufficient to prove that

$$\|\hat{\varphi}\|_2 = \|\varphi\|_2 , \quad \varphi \in \mathcal{S} .$$

Let ψ be the function defined by

$$\psi(x) = \overline{\varphi(-x)} , \quad x \in \mathbb{R}^n . \tag{3.15}$$

Then $\psi \in \mathcal{S}$ and

$$\hat{\psi}(\xi) = (2\pi)^{-n/2} \int_{\mathbb{R}^n} e^{-ix\cdot\xi} \overline{\varphi(-x)} dx$$

$$= (2\pi)^{-n/2} \int_{\mathbb{R}^n} e^{ix\cdot\xi} \overline{\varphi(x)} dx$$

$$= \overline{\hat{\varphi}(\xi)} . \tag{3.16}$$

Thus, by (3.15),

$$\|\varphi\|_2^2 = \int_{\mathbb{R}^n} \varphi(x)\overline{\varphi(x)} dx = \int_{\mathbb{R}^n} \varphi(x)\psi(-x) dx = (\varphi * \psi)(0) . \tag{3.17}$$

By Proposition 3.10, $\varphi * \psi \in \mathcal{S}$. Hence, by (3.16), Proposition 3.1 and the Fourier inversion formula,

$$(\varphi * \psi)(0) = (2\pi)^{-n/2} \int_{\mathbb{R}^n} (\varphi * \psi)\widehat{}(\xi) d\xi$$

$$= \int_{\mathbb{R}^n} \hat{\varphi}(\xi)\hat{\psi}(\xi) d\xi$$

$$= \int_{\mathbb{R}^n} \hat{\varphi}(\xi)\overline{\hat{\varphi}(\xi)} d\xi$$

$$= \|\hat{\varphi}\|_2^2 . \tag{3.18}$$

Therefore, by (3.17) and (3.18),

$$\|\varphi\|_2 = \|\hat{\varphi}\|_2 , \quad \varphi \in \mathcal{S} .$$

Remark 3.12.   The Plancherel Theorem states that the Fourier transform of a function in $L^2(\mathbb{R}^n)$ can be defined. If $f \in L^2(\mathbb{R}^n)$, then we shall denote its Fourier transform by $\hat{f}$ or $\mathcal{F}f$. The inverse of $\mathcal{F} : L^2(\mathbb{R}^n) \to L^2(\mathbb{R}^n)$ is of course denoted by $\mathcal{F}^{-1} : L^2(\mathbb{R}^n) \to L^2(\mathbb{R}^n)$.

Exercises

3.1.   Here is another elegant proof of Proposition 3.5.

(i) Let φ be the function defined on $\mathbb{R}$ by

$$\varphi(x) = e^{-\frac{x^2}{2}} , \quad x \in \mathbb{R} .$$

Let $y = \hat{\varphi}$. Prove that

$$y'(\xi) + \xi y(\xi) = 0 , \quad \xi \in \mathbb{R} .$$

(ii) Use the result in part (i) to prove that

$$\hat{\varphi}(\xi) = e^{-\frac{\xi^2}{2}} , \quad \xi \in \mathbb{R} .$$

3.2.   Let $\{\varphi_n\}$ be the sequence of functions defined on $\mathbb{R}$ by

$$\begin{cases} \varphi_0(x) = e^{-\frac{x^2}{2}} \\ \varphi_{n+1}(x) = x\varphi_n(x) - \varphi_n'(x) \end{cases}$$

for all $x \in \mathbb{R}$ and $n = 0, 1, 2, \ldots$. We call φ_n a *Hermite function of order* n.

(i) Prove that

$$\varphi_n(x) = (-1)^n e^{\frac{x^2}{2}} \left(\frac{d}{dx} \right)^n \left(e^{-x^2} \right)$$

for all $x \in \mathbb{R}$ and $n = 0, 1, 2, \ldots$.

(ii) Prove that $\hat{\varphi}_n = i^{-n} \varphi_n$ for $n = 0, 1, 2, \ldots$.

3.3.   Prove that if $f \in L^1(\mathbb{R}^n)$ and $\hat{f} \in L^1(\mathbb{R}^n)$, then $(\hat{f})^{\vee} = f$ a.e.

3.4.   Find a function f in $L^1(\mathbb{R}^n)$ such that $\hat{f}$ is not in $L^1(\mathbb{R}^n)$.

3.5.   Prove that $|x^\alpha| \leq |x|^{|\alpha|}$ for all $x \in \mathbb{R}^n$ and multi-indices α.

4. TEMPERED DISTRIBUTIONS

We only give the rudiments of the theory of tempered distributions in this chapter. More details on this subject will be introduced in later chapters as the need arises.

Definition 4.1. A sequence of functions $\{\varphi_j\}$ in the Schwartz space S is said to *converge to zero in* S (denoted by $\varphi_j \to 0$ in S) if for all multi-indices α and β, we have

$$\sup_{x \in \mathbb{R}^n} |x^\alpha (D^\beta \varphi_j)(x)| \to 0 \quad \text{as} \quad j \to \infty .$$

Definition 4.2. A linear functional T on S is called a *tempered distribution* if for any sequence $\{\varphi_j\}$ of functions in S converging to zero in S, we have

$$T(\varphi_j) \to 0 \quad \text{as} \quad j \to \infty .$$

Definition 4.3. Let f be a measurable function defined on $\mathbb{R}^n$ such that

$$\int_{\mathbb{R}^n} \frac{|f(x)|}{(1 + |x|)^N} dx < \infty$$

for some positive integer N. Then we call f a *tempered function*.

Proposition 4.4.   *Let f be a tempered function defined on $\mathbb{R}^n$. Then the linear functional T_f on S defined by*

$$T_f(\varphi) = \int_{\mathbb{R}^n} f(x)\varphi(x)dx , \quad \varphi \in S ,$$

is a tempered distribution.

Proof: Let N be a positive integer such that

$$\int_{\mathbb{R}^n} \frac{|f(x)|}{(1+|x|)^N} dx < \infty .$$

Then, for all functions $\varphi \in S$, the integral $\int_{\mathbb{R}^n} f(x)\varphi(x)dx$ exists. Indeed, we have

$$\int_{\mathbb{R}^n} |f(x)|\,|\varphi(x)|dx$$
$$= \int_{\mathbb{R}^n} \frac{|f(x)|}{(1+|x|)^N}(1+|x|)^N|\varphi(x)|dx$$
$$\leq \left(\int_{\mathbb{R}^n} \frac{|f(x)|}{(1+|x|)^N}dx \right) \sup_{x \in \mathbb{R}^n} \left\{ (1+|x|)^N|\varphi(x)| \right\} < \infty .$$

Let $\{\varphi_j\}$ be a sequence of functions in S converging to zero in S as $j \to \infty$. Then obviously,

$$\sup_{x \in \mathbb{R}^n} \left\{ (1+|x|)^N\,|\varphi_j(x)| \right\} \to 0 . \tag{4.1}$$

Since

$$|T_f(\varphi_j)| \leq \int_{\mathbb{R}^n} |f(x)|\,|\varphi_j(x)|\,dx$$
$$\leq \sup_{x \in \mathbb{R}^n} \left\{ (1+|x|)^N|\varphi_j(x)| \right\} \int_{\mathbb{R}^n} \frac{|f(x)|}{(1+|x|)^N}dx \tag{4.2}$$

for all j, it follows from (4.1) and (4.2) that $T_f(\varphi_j) \to 0$ as $j \to \infty$.

Proposition 4.5. *Let $f \in L^p(\mathbb{R}^n)$, $1 \le p \le \infty$. Then the linear functional T_f on S defined by*

$$T_f(\varphi) = \int_{\mathbb{R}^n} f(x)\varphi(x)dx , \quad \varphi \in S ,$$

is a tempered distribution.

Proof: Note that f is a tempered function. See Exercise 4.1.

Remark 4.6. It is customary to identify the tempered distribution T_f with the function f and to say that such tempered distributions are functions.

Definition 4.7. Let T be a tempered distribution. Then the Fourier transform of T is defined to be the linear functional $\hat{T}$ on S given by

$$\hat{T}(\varphi) = T(\hat{\varphi}) , \quad \varphi \in S .$$

Proposition 4.8. $\hat{T}$ *is also a tempered distribution.*

Proof: Let $\{\varphi_j\}$ be a sequence of functions in S converging to zero in S. We need only prove that the sequence $\{\hat{\varphi}_j\}$ also converges to zero in S. To this end, let α and β be any two multi-indices. Then, by Proposition 3.2,

$$\begin{aligned}
\sup_{\xi \in \mathbb{R}^n} \left| \xi^\alpha \left(D^\beta \hat{\varphi}_j \right)(\xi) \right| &= \sup_{\xi \in \mathbb{R}^n} \left| \xi^\alpha \widehat{\left((-x)^\beta \varphi_j \right)}(\xi) \right| \\
&= \sup_{\xi \in \mathbb{R}^n} \left| \widehat{\left\{ D^\alpha \left((-x)^\beta \varphi_j \right) \right\}}(\xi) \right| \\
&\le (2\pi)^{-n/2} \left\| D^\alpha \left((-x)^\beta \varphi_j \right) \right\|_1 .
\end{aligned}$$
$$(4.3)$$

Since $\varphi_j \to 0$ in S as $j \to \infty$, it follows that for any positive integer N, we have

$$\sup_{x \in \mathbb{R}^n} \left\{ (1 + |x|)^N \left| \left(D^\alpha \left((-x)^\beta \varphi_j \right) \right)(x) \right| \right\} \to 0 \qquad (4.4)$$

as $j \to \infty$. Since, for any positive integer N greater than n, we have

$$\| D^\alpha \left((-x)^\beta \varphi_j \right) \|_1$$

$$\leq \sup_{x \in \mathbb{R}^n} \left\{ (1 + |x|)^N \left| \left(D^\alpha \left((-x)^\beta \varphi_j \right) \right) (x) \right| \right\} \int_{\mathbb{R}^n} (1 + |x|)^{-N} dx$$

$$(4.5)$$

for all j. Hence, by (4.3), (4.4) and (4.5), we conclude that $\hat{\varphi}_j \to 0$ in $\mathcal{S}$ as $j \to \infty$.

Theorem 4.9. (The Fourier Inversion Formula) *Let T be a tempered distribution. Then*

$$\hat{\tilde{T}} = \tilde{T} \, ,$$

where $\tilde{T}$ is defined by

$$\tilde{T}(\varphi) = T(\tilde{\varphi}) \, , \quad \varphi \in \mathcal{S} \, .$$

Proof: Let $\varphi \in \mathcal{S}$. Then, by Definition 4.7 and the Fourier inversion formula for $\mathcal{S}$, we have

$$\hat{\hat{T}}(\varphi) = \hat{T}(\hat{\varphi}) = T(\hat{\hat{\varphi}}) = T(\tilde{\varphi}) = \tilde{T}(\varphi) \, .$$

Exercises

4.1.   Prove that any function in $L^p(\mathbb{R}^n), 1 \leq p \leq \infty$, is a tempered function.

4.2.   Let $\delta : \mathcal{S} \to \mathbb{C}$ be the mapping defined by

$$\delta(\varphi) = \varphi(0) \, , \quad \varphi \in \mathcal{S} \, .$$

(i) Prove that δ is a tempered distribution.

(ii) Prove that δ is not a tempered function. (See Remark 4.6.)

4.3. Let $f \in L^1(\mathbb{R}^n)$ and T be the tempered distribution which is equal to f. Prove that $\hat{T}$ is equal to $\hat{f}$.

4.4. Do Exercise 4.3 again for $f \in L^2(\mathbb{R}^n)$.

4.5. Find the Fourier transform of the tempered distribution δ defined in Exercise 4.2.

5. SYMBOLS, PSEUDO-DIFFERENTIAL OPERATORS AND ASYMPTOTIC EXPANSIONS

In this chapter we give the definition and the most elementary properties of a pseudo-differential operator and its symbol.

We begin by recalling that a linear partial differential operator $P(x, D)$ on $\mathbb{R}^n$ is given by

$$P(x, D) = \sum_{|\alpha| \leq m} a_\alpha(x) D^\alpha , \qquad (5.1)$$

where the coefficients $a_\alpha(x)$ are functions defined on $\mathbb{R}^n$. If we replace the D^α in (5.1) by the monomial ξ^α in $\mathbb{R}^n$, then we obtain the so-called *symbol*

$$P(x, \xi) = \sum_{|\alpha| \leq m} a_\alpha(x) \xi^\alpha \qquad (5.2)$$

of the operator (5.1). In order to get another representation of the operator $P(x, D)$, let us take any function φ in $\mathcal{S}$. Then, by (5.1), (5.2), Proposition 3.2 and the Fourier inversion formula for Schwartz functions, we have

$$(P(x, D)\varphi)(x) = \sum_{|\alpha| \le m} a_\alpha(x)(D^\alpha \varphi)(x)$$

$$= \sum_{|\alpha| \le m} a_\alpha(x)(\widehat{D^\alpha \varphi})^{\vee}(x)$$

$$= \sum_{|\alpha| \le m} a_\alpha(x)(\xi^\alpha \hat{\varphi})^{\vee}(x)$$

$$= \sum_{|\alpha| \le m} a_\alpha(x)(2\pi)^{-n/2} \int_{\mathbb{R}^n} e^{ix\cdot\xi} \xi^\alpha \hat{\varphi}(\xi) d\xi$$

$$= (2\pi)^{-n/2} \int_{\mathbb{R}^n} e^{ix\cdot\xi} \left(\sum_{|\alpha| \le m} a_\alpha(x)\xi^\alpha \right) \hat{\varphi}(\xi) d\xi$$

$$= (2\pi)^{-n/2} \int_{\mathbb{R}^n} e^{ix\cdot\xi} P(x, \xi)\hat{\varphi}(\xi) d\xi . \tag{5.3}$$

So we have represented the partial differential operator $P(x, D)$ in terms of its symbol by means of the Fourier transform. This representation immediately suggests that we can get operators more general than partial differential operators if we replace the symbol $P(x, \xi)$ by more general symbols $\sigma(x, \xi)$ which are no longer polynomials in ξ. The operators so obtained are called pseudo-differential operators. We shall do this in due course. Meanwhile, we should point out that in order to get a useful and tractable class of operators, it is necessary to impose certain conditions on the functions $\sigma(x, \xi)$. Many different sets of conditions have been proposed, resulting in many different classes of pseudo-differential operators. Our discussion in this book is restricted to the following class:

Definition 5.1. Let $m \in (-\infty, \infty)$. Then we define S^m to be the set of all functions $\sigma(x, \xi)$ in $C^\infty(\mathbb{R}^n \times \mathbb{R}^n)$ such that for any two multi-indices α and β, there is a positive constant $C_{\alpha,\beta}$, depending on α and β only, for which

$$\left| \left(D_x^\alpha D_\xi^\beta \sigma \right)(x, \xi) \right| \le C_{\alpha,\beta}(1 + |\xi|)^{m-|\beta|} , \quad x, \xi \in \mathbb{R}^n .$$

We call any function σ in $\cup_{m \in \mathbb{R}} S^m$ a *symbol*.

Definition 5.2. Let σ be a symbol. Then the *pseudo-differential operator* T_σ associated to σ is defined by

$$\left(T_\sigma \varphi\right)(x) = (2\pi)^{-n/2} \int_{\mathbb{R}^n} e^{ix \cdot \xi} \sigma(x, \xi) \hat{\varphi}(\xi) d\xi , \quad \varphi \in \mathcal{S} . \quad (5.4)$$

We give some examples.

Example 5.3. Let $P(x, D) = \sum_{|\alpha| \leq m} a_\alpha(x) D^\alpha$ be a linear partial differential operator on $\mathbb{R}^n$. If all the coefficients $a_\alpha(x)$ are C^∞ and have bounded derivatives of all orders, then the polynomial $P(x, \xi) = \sum_{|\alpha| \leq m} a_\alpha(x) \xi^\alpha$ is in S^m and hence $P(x, D)$ is a pseudo-differential operator. (See formulas (5.3) and (5.4).)

Proof: Let γ and δ be multi-indices. Then

$$\left|\left(D_x^\gamma D_\xi^\delta P\right)(x, \xi)\right| \leq \sum_{|\alpha| \leq m} C_{\alpha, \gamma} \left|\partial_\xi^\delta \xi^\alpha\right| \quad (5.5)$$

for all $x, \xi \in \mathbb{R}^n$, where $C_{\alpha, \gamma} = \sup_{x \in \mathbb{R}^n} \left|\left(D^\gamma a_\alpha\right)(x)\right|$. It can be shown easily that

$$\partial_\xi^\delta \xi^\alpha = \begin{cases} \delta! \dbinom{\alpha}{\delta} \xi^{\alpha - \delta} , & \delta \leq \alpha , \\ 0 , & \text{otherwise} , \end{cases} \quad (5.6)$$

for all $\xi \in \mathbb{R}^n$. The proof of (5.6) is left as an exercise. See Exercise 5.5. Hence, by (5.5), (5.6) and Exercise 3.5,

$$\left|\left(D_x^\gamma D_\xi^\delta P\right)(x, \xi)\right| \leq \sum_{|\alpha| \leq m} C_{\alpha, \gamma} \delta! \dbinom{\alpha}{\delta} |\xi|^{|\alpha| - |\delta|}$$

$$\leq C'_{\gamma, \delta} (1 + |\xi|)^{m - |\delta|}$$

for all $x, \xi \in \mathbb{R}^n$, where $C'_{\gamma, \delta} = \sum_{|\alpha| \leq m} C_{\alpha, \gamma} \delta! \dbinom{\alpha}{\delta}$.

Example 5.4. Let $\sigma(\xi) = (1+|\xi|^2)^{m/2}$, $-\infty < m < \infty$. Then $\sigma \in S^m$ and hence T_σ is a pseudo-differential operator. It is sometimes advantageous to denote T_σ by $(I - \Delta)^{m/2}$, where I is the identity operator and Δ the Laplacian, i.e., $\Delta = \sum\limits_{j=1}^{n} \dfrac{\partial^2}{\partial x_j^2}$.

Proof: We need only prove that for any $m \in (-\infty, \infty)$ and multi-index β, there exists a positive constant $C_{m,\beta}$ such that

$$\left|\left(D^\beta \sigma\right)(\xi)\right| \leq C_{m,\beta}(1 + |\xi|)^{m-|\beta|} \tag{5.7}$$

for all $\xi \in \mathbb{R}^n$. (5.7) is obviously true for the zero multi-index. Suppose that (5.7) is true for all $m \in (-\infty, \infty)$ and multi-indices β of length at most equal to l. Let γ be a multi-index of length $l + 1$. Then $D^\gamma = D^\beta D_j$ for some $j = 1, 2, \ldots, n$ and some multi-index β of length l. Hence

$$\left|(D^\gamma \sigma)(\xi)\right| = \left|(\partial^\beta \partial_j \sigma)(\xi)\right| = \left|(\partial^\beta \tau)(\xi)\right| \tag{5.8}$$

for all $\xi \in \mathbb{R}^n$, where

$$\tau(\xi) = m\xi_j \left(1 + |\xi|^2\right)^{\frac{m}{2}-1}$$

for all $\xi \in \mathbb{R}^n$. Therefore, by Leibnitz' formula,

$$\left(\partial^\beta \tau\right)(\xi) = m \sum_{\delta \leq \beta} \binom{\beta}{\delta} \left(\partial^\delta \xi_j\right) \partial^{\beta-\delta} \left\{ \left(1 + |\xi|^2\right)^{\frac{m}{2}-1} \right\} \tag{5.9}$$

for all $\xi \in \mathbb{R}^n$. Hence, by (5.9) and the induction hypothesis, there exists a positive constant $C_{m,\beta}$ such that

$$\left|(\partial^\beta \tau)(\xi)\right| \leq C_{m,\beta} \sum_{\delta \leq \beta} \binom{\beta}{\delta} (1 + |\xi|)^{1-|\delta|}(1 + |\xi|)^{m-2-|\beta|+|\delta|}$$

$$= C'_{m,\beta}(1 + |\xi|)^{m-|\gamma|}$$

for all $\xi \in \mathbb{R}^n$, where $C'_{m,\beta} = C_{m,\beta} \sum_{\delta \leq \beta} \binom{\beta}{\delta}$. Thus, by (5.8) and the principle of mathematical induction, the proof is complete.

We give two very simple properties of pseudo-differential operators.

Proposition 5.5. *Let σ and τ be two symbols such that $T_\sigma = T_\tau$. Then $\sigma = \tau$.*

We first prove a lemma.

Lemma 5.6. *Let f be a continuous tempered function such that*

$$T_f(\varphi) = 0 , \quad \varphi \in \mathcal{S} . \tag{5.10}$$

Then f is identically zero on $\mathbb{R}^n$.

Proof: Without loss of generality, we assume that f is real-valued. Suppose $f(x_0) \neq 0$ for some $x_0 \in \mathbb{R}^n$. Then there is an open ball $B(x_0, r)$ with centre x_0 and radius r on which f is strictly positive (or negative). Choose a nonzero function $\varphi_0 \in C_0^\infty(\mathbb{R}^n)$ such that $\varphi_0(x) \geq 0$ for all $x \in \mathbb{R}^n$ and $\text{supp}(\varphi_0) \subset B(x_0, r)$. Such a function exists by Exercise 2.2. It is clear that $\varphi_0 \in \mathcal{S}$ and

$$T_f(\varphi_0) = \int_{\mathbb{R}^n} f(x)\varphi_0(x)dx$$

is strictly positive (or negative). This contradicts (5.10).

Proof of Proposition 5.5. By hypothesis and Definition 5.2, we have

$$\int_{\mathbb{R}^n} e^{ix\cdot\xi} \{\sigma(x,\xi) - \tau(x,\xi)\} \hat{\varphi}(\xi)d\xi = 0 , \quad \varphi \in \mathcal{S} .$$

Since $\mathcal{F} : \mathcal{S} \to \mathcal{S}$ is one-to-one and onto by the Fourier inversion formula, it follows that

$$\int_{\mathbb{R}^n} e^{ix\cdot\xi} \{\sigma(x,\xi) - \tau(x,\xi)\}\varphi(\xi)d\xi = 0 , \quad \varphi \in \mathcal{S} .$$

Now, for any fixed $x \in \mathbb{R}^n$, $e^{ix \cdot \xi}\{\sigma(x,\xi) - \tau(x,\xi)\}$ is a continuous tempered function in the variable ξ. This follows easily from the definition of a symbol in Definition 5.1. Hence, by Lemma 5.6, $e^{ix \cdot \xi}\{\sigma(x,\xi) - \tau(x,\xi)\} = 0$ for all $\xi \in \mathbb{R}^n$. Since x is fixed but arbitrary, it follows that $\sigma(x,\xi) - \tau(x,\xi) = 0$ for all $x, \xi \in \mathbb{R}^n$. This proves that $\sigma = \tau$.

Proposition 5.7. *Let σ be a symbol. Then T_σ maps the Schwartz space S into itself.*

Proof: Let $\varphi \in S$. Then, for any two multi-indices α and β, we need only prove that

$$\sup_{x \in \mathbb{R}^n} \left| x^\alpha \left(D^\beta (T_\sigma \varphi) \right)(x) \right| < \infty .$$

But, using integration by parts and Leibnitz' formula, we have

$$x^\alpha \left(D^\beta (T_\sigma \varphi) \right)(x) = x^\alpha (2\pi)^{-n/2} \int_{\mathbb{R}^n} D_x^\beta \left\{ e^{ix \cdot \xi} \sigma(x,\xi) \right\} \hat{\varphi}(\xi) d\xi$$

$$= x^\alpha (2\pi)^{-n/2} \int_{\mathbb{R}^n} \sum_{\gamma \leq \beta} \binom{\beta}{\gamma} \xi^\gamma e^{ix \cdot \xi}$$
$$\left(D_x^{\beta - \gamma} \sigma \right)(x,\xi) \hat{\varphi}(\xi) d\xi$$

$$= (2\pi)^{-n/2} \int_{\mathbb{R}^n} \sum_{\gamma \leq \beta} \binom{\beta}{\gamma} \xi^\gamma$$
$$\left(D_\xi^\alpha e^{ix \cdot \xi} \right) \left(D_x^{\beta - \gamma} \sigma \right)(x,\xi) \hat{\varphi}(\xi) d\xi$$

$$= (2\pi)^{-n/2} (-1)^{|\alpha|} \int_{\mathbb{R}^n} \sum_{\gamma \leq \beta} \binom{\beta}{\gamma} e^{ix \cdot \xi}$$
$$D_\xi^\alpha \left\{ \left(D_x^{\beta - \gamma} \sigma \right)(x,\xi) \xi^\gamma \hat{\varphi}(\xi) \right\} d\xi$$

$$= (2\pi)^{-n/2} (-1)^{|\alpha|} \int_{\mathbb{R}^n} \sum_{\gamma \leq \beta} \sum_{\delta \leq \alpha} \binom{\beta}{\gamma} \binom{\alpha}{\delta} e^{ix \cdot \xi}$$
$$\left(D_\xi^{\alpha - \delta} D_x^{\beta - \gamma} \sigma \right)(x,\xi) D_\xi^\delta \left(\xi^\gamma \hat{\varphi}(\xi) \right) d\xi .$$
$$(5.11)$$

Using (5.11) and the fact that σ is a symbol, say $\sigma \in S^m$, we can find positive constants $C_{\alpha,\beta,\gamma,\delta}$, depending on α, β, γ and δ only, such that

$$\sup_{x \in \mathbb{R}^n} \left| x^\alpha \left(D^\beta (T_\sigma \varphi) \right)(x) \right|$$

$$\leq (2\pi)^{-n/2} \sum_{\gamma \leq \beta} \sum_{\delta \leq \alpha} \binom{\beta}{\gamma} \binom{\alpha}{\delta} C_{\alpha,\beta,\gamma,\delta} \int_{\mathbb{R}^n} (1 + |\xi|)^{m - |\alpha| + |\delta|}$$

$$\left| D_\xi^\delta \left(\xi^\gamma \hat{\varphi}(\xi) \right) \right| d\xi .$$

(5.12)

Since $\varphi \in \mathcal{S}$, it follows from (5.12) that

$$\sup_{x \in \mathbb{R}^n} \left| x^\alpha \left(D^\beta (T_\sigma \varphi) \right)(x) \right| < \infty ,$$

and hence the proof is complete if we can justify the interchange of the order of differentiation and integration in (5.11). But, using the same argument as in the derivation of (5.12), we see that the last integral in (5.11) is absolutely convergent. This completes the proof.

Remark 5.8. In general, a pseudo-differential operator does not map $C_0^\infty(\mathbb{R}^n)$ into $C_0^\infty(\mathbb{R}^n)$. For a way of showing this, see Exercise 5.4.

An important notion in the theory of pseudo-differential operators is the asymptotic expansion of a symbol.

Definition 5.9. Let $\sigma \in S^m$. Suppose we can find $\sigma_j \in S^{m_j}$, where

$$m = m_0 > m_1 > m_2 > \ldots > m_j \to -\infty , \quad j \to \infty ,$$

such that

$$\sigma - \sum_{j=0}^{N-1} \sigma_j \in S^{m_N}$$

(5.13)

for every positive integer N. Then we call $\sum_{j=0}^{\infty} \sigma_j$ an *asymptotic expansion* of σ and we write

$$\sigma \sim \sum_{j=0}^{\infty} \sigma_j .$$

An important result in this connection is the following theorem:

Theorem 5.10. *Let $m_0 > m_1 > m_2 > \ldots > m_j \to -\infty$ as $j \to \infty$. Suppose $\sigma_j \in S^{m_j}$. Then there exists a symbol $\sigma \in S^{m_0}$ such that $\sigma \sim \sum_{j=0}^{\infty} \sigma_j$. Moreover, if τ is another symbol with the same asymptotic expansion, then $\sigma - \tau \in \bigcap_{m \in \mathbb{R}} S^m$.*

Proof: Let $\psi \in C^{\infty}(\mathbb{R}^n)$ be such that $0 \le \psi(\xi) \le 1$ for $\xi \in \mathbb{R}^n, \psi(\xi) = 0$ for $|\xi| \le 1$ and $\psi(\xi) = 1$ for $|\xi| \ge 2$. That such a function exists will be proved at the end of this chapter. Let $\{\varepsilon_j\}$ be a sequence of positive numbers such that $1 > \varepsilon_0 > \varepsilon_1 > \varepsilon_2 > \ldots > \varepsilon_j \to 0$ as $j \to \infty$. Define the function σ on $\mathbb{R}^n \times \mathbb{R}^n$ by

$$\sigma(x, \xi) = \sum_{j=0}^{\infty} \psi(\varepsilon_j \xi) \sigma_j(x, \xi) , \quad x, \xi \in \mathbb{R}^n . \tag{5.14}$$

Note that for each $(x_0, \xi_0) \in \mathbb{R}^n \times \mathbb{R}^n$, there exists a neighborhood U of (x_0, ξ_0) and a positive integer N such that $\psi(\varepsilon_j \xi) \sigma_j(x, \xi) = 0$ for all $(x, \xi) \in U$ and $j > N$. Hence $\sigma \in C^{\infty}(\mathbb{R}^n \times \mathbb{R}^n)$. Furthermore, for any $\varepsilon \in (0, 1]$ and nonzero multi-index $\alpha, \psi(\varepsilon \xi) = 0$ if $|\xi| \le \frac{1}{\varepsilon}, \psi(\varepsilon \xi) = 1$ if $|\xi| \ge \frac{2}{\varepsilon}, \partial_{\xi}^{\alpha}\{\psi(\varepsilon \xi)\} = \varepsilon^{|\alpha|}(\partial^{\alpha}\psi)(\varepsilon \xi) = 0$ if $|\xi| \le \frac{1}{\varepsilon}$ or $|\xi| \ge \frac{2}{\varepsilon}$ and $|\partial_{\xi}^{\alpha}\{\psi(\varepsilon \xi)\}| \le C_{\alpha} \varepsilon^{|\alpha|}$ for all $\xi \in \mathbb{R}^n$, where $C_{\alpha} = \sup_{\xi \in \mathbb{R}^n} |(\partial^{\alpha}\psi)(\xi)|$. If $\frac{1}{\varepsilon} \le |\xi| \le \frac{2}{\varepsilon}$, then $\varepsilon \le \frac{2}{|\xi|} \le \frac{4}{1 + |\xi|}$. Hence, for any nonzero multi-index α, we have

$$|\partial_{\xi}^{\alpha}\{\psi(\varepsilon \xi)\}| \le C_{\alpha} 4^{|\alpha|}(1 + |\xi|)^{-|\alpha|} = C_{\alpha}'(1 + |\xi|)^{-|\alpha|} , \quad \xi \in \mathbb{R}^n , \tag{5.15}$$

where $C'_\alpha = C_\alpha 4^{|\alpha|}$. It is obvious that (5.15) is also true for the zero multi-index α. Now, using (5.15), Leibnitz' formula and the fact that $\sigma_j \in S^{m_j}$, we can find positive constants $C_{\alpha,\gamma}$ and $C_{j,\beta,\gamma}$ such that

$$
\begin{aligned}
&\left| D_\xi^\alpha D_x^\beta \left\{ \psi(\varepsilon\xi)\sigma_j(x,\xi) \right\} \right| \\
&= \left| D_\xi^\alpha \{ \psi(\varepsilon\xi) \left(D_x^\beta \sigma_j \right)(x,\xi) \} \right| \\
&= \left| \sum_{\gamma \le \alpha} \binom{\alpha}{\gamma} \left(D_\xi^{\alpha-\gamma} \{ \psi(\varepsilon\xi) \} \right) \left(D_\xi^\gamma D_x^\beta \sigma_j \right)(x,\xi) \right| \\
&\le \sum_{\gamma \le \alpha} \binom{\alpha}{\gamma} C_{\alpha,\gamma} (1+|\xi|)^{-|\alpha|+|\gamma|} C_{j,\beta,\gamma} (1+|\xi|)^{m_j-|\gamma|} \\
&= \sum_{\gamma \le \alpha} \binom{\alpha}{\gamma} C_{\alpha,\gamma} C_{j,\beta,\gamma} (1+|\xi|)^{m_j-|\alpha|} \\
&= C_{j,\alpha,\beta} (1+|\xi|)^{-1} (1+|\xi|)^{m_j+1-|\alpha|}
\end{aligned}
\tag{5.16}
$$

for all $x,\xi \in \mathbb{R}^n$, where $C_{j,\alpha,\beta} = \sum_{\gamma \le \alpha} \binom{\alpha}{\gamma} C_{\alpha,\gamma} C_{j,\beta,\gamma}$. Now we choose ε_j such that

$$
C_{j,\alpha,\beta} \varepsilon_j \le 2^{-j}
\tag{5.17}
$$

for all multi-indices α and β such that $|\alpha+\beta| \le j$. By the definition of ψ, we have

$$
\psi(\varepsilon_j \xi) = 0
\tag{5.18}
$$

whenever $1 + |\xi| \le \varepsilon_j^{-1}$. Hence, by (5.16), (5.17) and (5.18),

$$
\left| D_\xi^\alpha D_x^\beta \{ \psi(\varepsilon_j \xi)\sigma_j(x,\xi) \} \right| \le 2^{-j}(1+|\xi|)^{m_j+1-|\alpha|}
\tag{5.19}
$$

whenever $x,\xi \in \mathbb{R}^n$ and $|\alpha+\beta| \le j$. Now, for any two multi-indices α_0 and β_0, we take j_0 so large that $j_0 \ge |\alpha_0+\beta_0|$ and $m_{j_0}+1 \le m_0$. Write

$$
\begin{aligned}
\sigma(x,\xi) &= \sum_{j=0}^{j_0-1} \psi(\varepsilon_j \xi)\sigma_j(x,\xi) + \sum_{j=j_0}^\infty \psi(\varepsilon_j \xi)\sigma_j(x,\xi) \\
&= I(x,\xi) + J(x,\xi) .
\end{aligned}
\tag{5.20}
$$

Since $I(x,\xi)$ is a finite sum, it follows that $I \in S^{m_0}$. By (5.19),

$$
\left|(D_\xi^{\alpha_0} D_x^{\beta_0} J)(x,\xi)\right| \leq \sum_{j=j_0}^\infty \left|D_\xi^{\alpha_0} D_x^{\beta_0} \{\psi(\varepsilon_j\xi)\sigma_j(x,\xi)\}\right|
$$

$$
\leq \sum_{j=j_0}^\infty 2^{-j}(1+|\xi|)^{m_j+1-|\alpha_0|}
$$

$$
\leq \sum_{j=j_0}^\infty 2^{-j}(1+|\xi|)^{m_0-|\alpha_0|}
$$

$$
= 2^{-j_0+1}(1+|\xi|)^{m_0-|\alpha_0|} . \tag{5.21}
$$

So J is also in S^{m_0}. Hence, by (5.20), $\sigma \in S^{m_0}$. We need to verify condition (5.13). To do this, we write

$$
\sigma(x,\xi) - \sum_{j=0}^{N-1} \sigma_j(x,\xi) = \sum_{j=0}^\infty \psi(\varepsilon_j\xi)\sigma_j(x,\xi) - \sum_{j=0}^{N-1} \sigma_j(x,\xi)
$$

$$
= \sum_{j=0}^{N-1} \{\psi(\varepsilon_j\xi) - 1\} \sigma_j(x,\xi)
$$

$$
+ \sum_{j=N}^\infty \psi(\varepsilon_j\xi)\sigma_j(x,\xi) .
$$

As before, we can show that $\displaystyle\sum_{j=N}^\infty \psi(\varepsilon_j\xi)\sigma_j(x,\xi) \in S^{m_N}$. Since $\psi(\varepsilon_j\xi) - 1 = 0$ for $j \leq N-1$ if $|\xi| \geq \dfrac{2}{\varepsilon_{N-1}}$, it follows from Exercise 5.3 that

$$
\sum_{j=0}^{N-1} \{\psi(\varepsilon_j\xi) - 1\} \sigma_j(x,\xi) \in \underset{m\in\mathbb{R}}{\cap} S^m
$$

and consequently $\sigma - \displaystyle\sum_{j=0}^{N-1} \sigma_j \in S^{m_N}$. Finally, if τ is another symbol such that $\tau \sim \sum_{j=0}^{\infty} \sigma_j$, then

$$\sigma - \tau = \left[\sigma - \sum_{j=0}^{N-1} \sigma_j \right] - \left[\tau - \sum_{j=0}^{N-1} \sigma_j \right] \in S^{m_N}$$

for every positive integer N. Since $m_N \to -\infty$ as $N \to \infty$, it follows that $\sigma - \tau \in \bigcap_{m \in \mathbb{R}} S^m$. This completes the proof of the theorem.

We have used the following result in the proof of Theorem 5.10.

Proposition 5.11. *There exists a function* $\psi \in C^{\infty}(\mathbb{R}^n)$ *such that* $0 \leq \psi(\xi) \leq 1$ *for* $\xi \in \mathbb{R}^n, \psi(\xi) = 0$ *for* $|\xi| \leq 1$ *and* $\psi(\xi) = 1$ *for* $|\xi| \geq 2$.

Proof: We need only construct a function $\varphi_0 \in C_0^{\infty}(\mathbb{R}^n)$ such that $0 \leq \varphi_0(\xi) \leq 1$ for $\xi \in \mathbb{R}^n, \varphi_0(\xi) = 1$ for $|\xi| \leq 1$ and $\varphi_0(\xi) = 0$ for $|\xi| \geq 2$. For then the function $\psi = 1 - \varphi_0$ will satisfy all the conditions of Proposition 5.11. To construct φ_0, let f be any continuous function on $\mathbb{R}^n$ such that $0 \leq f(t) \leq 1$ for $t \in \mathbb{R}^n, f(t) = 1$ for $|t| \leq \frac{3}{2}$ and $f(t) = 0$ for $|t| \geq \frac{7}{4}$. Let $\varphi \in C_0^{\infty}(\mathbb{R}^n)$ be a real-valued and nonnegative function such that $\varphi(s) = 0$ for $|s| \geq \frac{1}{4}$ and

$$\int_{|s| \leq \frac{1}{4}} \varphi(s) ds = 1 . \tag{5.22}$$

That such a function φ exists is an immediate consequence of Exercise 2.2. Let $\varphi_0 = f * \varphi$. Then, by Propositions 2.6 and 2.7, $\varphi_0 \in C_0^{\infty}(\mathbb{R}^n)$. Furthermore, by Proposition 2.7 and the location of the supports of f and φ, we see that $\varphi_0(t) = 0$ for $|t| \geq 2$. Finally, for $|t| \leq 1$,

$$\varphi_0(t) = \int_{\mathbb{R}^n} f(t - s) \varphi(s) ds = \int_{|s| \leq \frac{1}{4}} f(t - s) \varphi(s) ds . \tag{5.23}$$

Since, for $|t| \leq 1$ and $|s| \leq \dfrac{1}{4}$, we have $|s - t| \leq \dfrac{5}{4}$ and hence $f(t - s) = 1$. Therefore, by (5.22) and (5.23), $\varphi_0(t) = 1$ for $|t| \leq 1$. Again, by (5.22) and (5.23), we can prove easily that $0 \leq \varphi_0(t) \leq 1$ for all t in $\mathbb{R}^n$. This completes the proof of the proposition.

Remark 5.12.   For another proof of Proposition 5.11, see Exercise 5.6.

Exercises

5.1.   Prove that if $\sigma \in S^{m_1}$ and $\tau \in S^{m_2}$, then $\sigma \tau \in S^{m_1 + m_2}$.

5.2.   Prove that if $\sigma \in S^m$, then $D_x^\alpha D_\xi^\beta \sigma \in S^{m - |\beta|}$ for all multi-indices α and β.

5.3.   Let σ be any symbol and φ any function in $\mathcal{S}$. Prove that the function τ defined by

$$\tau(x, \xi) = \sigma(x, \xi)\varphi(\xi) , \quad x, \xi \in \mathbb{R}^n ,$$

is a symbol in $\bigcap_{m \in \mathbb{R}} S^m$.

5.4.   Let σ be the symbol defined by

$$\sigma(\xi) = e^{-\frac{|\xi|^2}{2}}$$

for all $\xi \in \mathbb{R}^n$. Show that the pseudo-differential operator T_σ does not map $C_0^\infty(\mathbb{R}^n)$ into $C_0^\infty(\mathbb{R}^n)$.

5.5.   Let α and δ be any two multi-indices. Prove that for all $\xi \in \mathbb{R}^n$,

$$\partial_\xi^\delta \xi^\alpha = \begin{cases} \delta! \begin{pmatrix} \alpha \\ \delta \end{pmatrix} \xi^{\alpha - \delta} , & \delta \leq \alpha \\ 0 , & \text{otherwise} . \end{cases}$$

5.6. (i)   Let $\varphi \in C_0^\infty(\mathbb{R})$ be such that $\varphi(x) \geq 0$ for $x \in \mathbb{R}, \varphi(x) = 0$ for $x \notin [-2, -1]$, and

$$\int_{-\infty}^{\infty} \varphi(x)dx = 1 .$$

Define a function χ on $(-\infty, 0]$ by

$$\chi(\xi) = \int_{-\infty}^{\xi} \varphi(x)dx , \quad \xi \in (-\infty, 0] .$$

Prove that χ is infinitely differentiable on $(-\infty, 0), 0 \leq \chi(\xi) \leq 1$ for $\xi \in (-\infty, 0], \chi(\xi) = 0$ for $\xi \leq -2$ and $\chi(\xi) = 1$ for $\xi \in [-1, 0]$.

(ii) Let $\varphi \in C_0^\infty(\mathbb{R})$ be such that $\varphi(x) \geq 0$ for $x \in \mathbb{R}, \varphi(x) = 0$ for $x \notin [1, 2]$ and

$$\int_{-\infty}^{\infty} \varphi(x)dx = 1 .$$

Define a function χ on $[0, \infty)$ by

$$\chi(\xi) = \int_{\xi}^{\infty} \varphi(x)dx , \quad \xi \in [0, \infty).$$

Prove that χ is infinitely differentiable on $(0, \infty), 0 \leq \chi(\xi) \leq 1$ for $\xi \in [0, \infty), \chi(\xi) = 0$ for $\xi \geq 2$ and $\chi(\xi) = 1$ for $\xi \in [0, 1]$.

(iii) Construct a function $\psi \in C^\infty(\mathbb{R}^n)$ such that $0 \leq \psi(\xi) \leq 1$ for $\xi \in \mathbb{R}^n, \psi(\xi) = 0$ for $|\xi| \leq 1$ and $\psi(\xi) = 1$ for $|\xi| \geq 2$.

6. A PARTITION OF UNITY AND TAYLOR'S FORMULA

It is convenient to devote a chapter to several technical results which will be of particular importance for us in the next two chapters. In Theorem 6.1 we construct a partition of unity. Then we use this partition of unity to decompose a symbol $\sigma(x,\xi)$ into a family $\{\sigma_k(x,\xi)\}$ of symbols with compact supports in the ξ variables. We are able to obtain good estimates on the partial Fourier transforms (with respect to the ξ variables) of all the symbols $\sigma_k(x,\xi)$. The precise estimates are given in Theorem 6.2. In Theorem 6.3 we prove a multi-dimensional version of Taylor's formula with integral remainder.

We begin with the construction of a partition of unity.

Theorem 6.1. *There is a sequence* $\{\varphi_k\}_{k=0}^{\infty}$ *of functions in* $C_0^{\infty}(\mathbb{R}^n)$ *such that*

(i) $0 \leq \varphi_k(\xi) \leq 1$, $\xi \in \mathbb{R}^n$, $k = 0, 1, 2, \ldots$,

(ii) $\displaystyle\sum_{k=0}^{\infty} \varphi_k(\xi) = 1$, $\xi \in \mathbb{R}^n$,

(iii) *at each* $\xi \in \mathbb{R}^n$*, at least one and at most three of the* φ_k*'s are nonzero,*

(iv) $\operatorname{supp}(\varphi_0) \subseteq \{\xi \in \mathbb{R}^n : |\xi| \leq 2\}$,

(v) $\operatorname{supp}(\varphi_k) \subseteq \{\xi \in \mathbb{R}^n \colon 2^{k-2} \leq |\xi| \leq 2^{k+1}\}$,　$k = 1, 2, \ldots$,

(vi) *for each multi-index α, there is a constant $A_\alpha > 0$ such that*

$$\sup_{\xi \in \mathbb{R}^n} |(\partial^\alpha \varphi_k)(\xi)| \leq A_\alpha 2^{-k|\alpha|} , \quad k = 0, 1, 2, \ldots .$$

Proof:　We pick ψ_0 to be any function in $C_0^\infty(\mathbb{R}^n)$ such that $0 \leq \psi_0(\xi) \leq 1$ for all $\xi \in \mathbb{R}^n, \psi_0(\xi) = 1$ for $|\xi| \leq 1$ and $\psi_0(\xi) = 0$ for $|\xi| > 2$. For the existence of such a function, see Proposition 5.11 and Remark 5.12. Let ψ be any other function in $C_0^\infty(\mathbb{R}^n)$ such that $0 \leq \psi(\xi) \leq 1$ for all $\xi \in \mathbb{R}^n$, $\psi(\xi) = 1$ for $1 \leq |\xi| \leq 2$, and $\psi(\xi) = 0$ for $|\xi| < \frac{1}{2}$ or $|\xi| > 4$. The proof that such a function ψ exists is left as an exercise. See Exercise 6.1. We give in Figure 1 the graph of such a function ψ in one dimension.

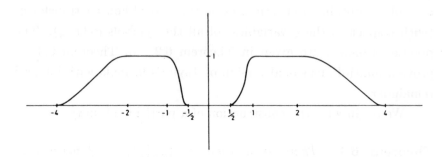

Fig. 1.

For $k = 1, 2, \ldots$, define ψ_k by

$$\psi_k(\xi) = \psi\left(\frac{\xi}{2^{k-1}}\right) , \quad \xi \in \mathbb{R}^n$$

and then define Ψ by

$$\Psi(\xi) = \sum_{k=0}^{\infty} \psi_k(\xi) , \quad \xi \in \mathbb{R}^n .$$

Obviously, we have

$$\text{supp}(\psi_0) \subseteq \{\xi \in \mathbb{R}^n : |\xi| \le 2\}$$

and

$$\text{supp}(\psi_k) \subseteq \left\{\xi \in \mathbb{R}^n : 2^{k-2} \le |\xi| \le 2^{k+1}\right\}$$

for $k = 1, 2, \ldots$. For each $\xi \in \mathbb{R}^n$, the series defining $\Psi(\xi)$ contains at most three nonzero consecutive terms. This implies that for each $\xi_0 \in \mathbb{R}^n$, there exists a neighborhood U of ξ_0 and a positive integer N such that $\psi_k(\xi) = 0$ for all $\xi \in U$ and $k > N$. Hence $\Psi \in C^\infty(\mathbb{R}^n)$. It is easy to see that for each $\xi \in \mathbb{R}^n$, the series defining $\Psi(\xi)$ contains at least one nonzero term. Hence $\Psi(\xi) \ne 0$ for all $\xi \in \mathbb{R}^n$. Now, for $k = 0, 1, 2, \ldots$, we define φ_k by

$$\varphi_k(\xi) = \frac{\psi_k(\xi)}{\Psi(\xi)}, \quad \xi \in \mathbb{R}^n .$$

It is easy to see that $\varphi_k \in C_0^\infty(\mathbb{R}^n)$ for $k = 0, 1, 2, \ldots$ and the sequence $\{\varphi_k\}_{k=0}^\infty$ satisfies the first five conditions. The last condition is trivially true for φ_0. Hence it remains to prove that the last condition is true for $\varphi_k, k = 1, 2, \ldots$. To do this, we note that for each multi-index α, we have, by Leibnitz' formula,

$$(\partial^\alpha \varphi_k)(\xi) = \sum_{\beta \le \alpha} \binom{\alpha}{\beta} \left\{\partial^\beta\left(\frac{1}{\Psi}\right)(\xi)\right\}(\partial^{\alpha-\beta}\psi_k)(\xi)$$

$$= \sum_{\beta \le \alpha} \binom{\alpha}{\beta} \left\{\partial^\beta\left(\frac{1}{\Psi}\right)(\xi)\right\}$$

$$\left(\partial^{\alpha-\beta}\psi\right)\left(\frac{\xi}{2^{k-1}}\right) 2^{-(k-1)|\alpha-\beta|} . \tag{6.1}$$

Now, for each multi-index β, by formula (0.4),

$$\partial^\beta\left(\frac{1}{\Psi}\right) = \sum_{\beta^{(1)} + \cdots + \beta^{(l)} = \beta} C_{\beta^{(1)}, \ldots, \beta^{(l)}} \frac{\left(\partial^{\beta^{(1)}}\Psi\right) \cdots \left(\partial^{\beta^{(l)}}\Psi\right)}{\Psi^{l+1}} , \tag{6.2}$$

where $C_{\beta^{(1)},\ldots,\beta^{(l)}}$ is a constant and the sum is taken over all possible multi-indices $\beta^{(1)}, \ldots, \beta^{(l)}$ which partition β. Next, for every multi-index γ, there exists a constant $C_\gamma > 0$ such that

$$|(\partial^\gamma \Psi)(\xi)| \leq C_\gamma 2^{-k|\gamma|} \tag{6.3}$$

for all $\xi \in \text{supp}(\varphi_k)$ and $k = 0, 1, 2, \ldots$. Let us assume (6.3) for a moment. Then, by (6.2) and (6.3), there exist positive constants $C_{\beta^{(1)}}, \ldots, C_{\beta^{(l)}}$ such that

$$\left| \partial^\beta \left(\frac{1}{\Psi} \right)(\xi) \right| \leq \sum_{\beta^{(1)}+\cdots+\beta^{(l)}=\beta} \left| C_{\beta^{(1)},\ldots,\beta^{(l)}} \right| \frac{C_{\beta^{(1)}} \cdots C_{\beta^{(l)}} 2^{-k|\beta|}}{|\Psi(\xi)|^{l+1}}$$

$$\leq C_\beta' 2^{-k|\beta|} \tag{6.4}$$

for all $\xi \in \text{supp}(\varphi_k)$, where

$$C_\beta' = \sum_{\beta^{(1)}+\cdots+\beta^{(l)}=\beta} \left| C_{\beta^{(1)},\ldots,\beta^{(l)}} \right| \left| C_{\beta^{(1)}} \cdots C_{\beta^{(l)}} \right| .$$

Hence, by (6.1) and (6.4), there exist positive constants $C_{\alpha,\beta}$ such that

$$\left| (\partial^\alpha \varphi_k)(\xi) \right| \leq \sum_{\beta \leq \alpha} \binom{\alpha}{\beta} C_\beta' 2^{-k|\beta|} C_{\alpha,\beta} 2^{-(k-1)|\alpha-\beta|}$$

$$= A_\alpha 2^{-k|\alpha|}$$

for all $k = 0, 1, 2, \ldots$ and all $\xi \in \mathbb{R}^n$, where

$$A_\alpha = \sum_{\beta \leq \alpha} \binom{\alpha}{\beta} C_\beta' C_{\alpha,\beta} 2^{|\alpha-\beta|} .$$

It remains to prove (6.3). To do this, we consider three cases.

Case 1: Suppose $k = 0$. Then for all $\xi \in \text{supp}(\varphi_0), \xi$ is in $\text{supp}(\psi_j)$ for some $j = 0, 1, 2$ and hence

$$\Psi(\xi) = \psi_0(\xi) + \sum_{j=1}^{2} \psi_j(\xi) .$$

Therefore

$$(\partial^\gamma \Psi)(\xi) = (\partial^\gamma \psi_0)(\xi) + \sum_{j=1}^{2} (\partial^\gamma \psi) \left(\frac{\xi}{2^{j-1}}\right) 2^{-(j-1)|\gamma|}$$

for all $\xi \in \text{supp}(\varphi_0)$. Thus there is a constant $C_\gamma > 0$ such that

$$\left|\left(\partial^\gamma \Psi\right)(\xi)\right| \leq C_\gamma , \quad \xi \in \text{supp}(\varphi_0) . \tag{6.5}$$

Case 2: Suppose $k = 1$. Then for all $\xi \in \text{supp}(\varphi_1), \xi$ is in $\text{supp}(\psi_j)$ for some $j = 0, 1, 2, 3$ and hence

$$\Psi(\xi) = \psi_0(\xi) + \sum_{j=1}^{3} \psi_j(\xi) .$$

Therefore

$$(\partial^\gamma \Psi)(\xi) = (\partial^\gamma \psi_0)(\xi) + \sum_{j=1}^{3} (\partial^\gamma \psi) \left(\frac{\xi}{2^{j-1}}\right) 2^{-(j-1)|\gamma|}$$

for all $\xi \in \text{supp}(\varphi_1)$. Thus there is a constant $C'_\gamma > 0$ such that

$$\begin{aligned}
|(\partial^\gamma \Psi)(\xi)| &\leq C'_\gamma \left[2^{|\gamma|} + 1 + 2^{-|\gamma|} + 2^{-2|\gamma|}\right] \\
&= C'_\gamma \left[2^{2|\gamma|} + 2^{|\gamma|} + 1 + 2^{-|\gamma|}\right] 2^{-|\gamma|} \tag{6.6}
\end{aligned}$$

for all $\xi \in \text{supp}(\varphi_1)$.

Case 3: Suppose $k \geq 2$. Then for all $\xi \in \text{supp}(\varphi_k), \xi$ is in $\text{supp}(\psi_j)$ for some $j = k - 2, k - 1, k, k + 1, k + 2$ and hence

$$\Psi(\xi) = \sum_{j=k-2}^{k+2} \psi_j(\xi) .$$

Therefore

$$(\partial^\gamma \Psi)(\xi) = \sum_{j=k-2}^{k+2} (\partial^\gamma \psi) \left(\frac{\xi}{2^{j-1}} \right) 2^{-(j-1)|\gamma|}$$

for all $\xi \in \operatorname{supp}(\varphi_k)$. Thus there is a constant $C_\gamma'' > 0$ such that

$$|(\partial^\gamma \Psi)(\xi)| \le C_\gamma'' \left[2^{-(k-3)|\gamma|} + 2^{-(k-2)|\gamma|} + 2^{-(k-1)|\gamma|} \right.$$

$$\left. + 2^{-k|\gamma|} + 2^{-(k+1)|\gamma|} \right]$$

$$= C_\gamma'' \left[2^{3|\gamma|} + 2^{2|\gamma|} + 2^{|\gamma|} + 1 + 2^{-|\gamma|} \right] 2^{-k|\gamma|}$$

$$\tag{6.7}$$

for all $\xi \in \operatorname{supp}(\varphi_k)$. Hence, by (6.5), (6.6) and (6.7), (6.3) follows and the proof of Theorem 6.1 is complete.

Let $\sigma \in S^m$. For $k = 0, 1, 2, \ldots$, we write

$$\sigma_k(x, \xi) = \sigma(x, \xi)\varphi_k(\xi) \tag{6.8}$$

for all $x, \xi \in \mathbb{R}^n$ and

$$K_k(x, z) = (2\pi)^{-n/2} \int_{\mathbb{R}^n} e^{iz \cdot \xi} \sigma_k(x, \xi) d\xi \tag{6.9}$$

for all $x, z \in \mathbb{R}^n$, where $\{\varphi_k\}$ is the partition of unity constructed in Theorem 6.1.

Theorem 6.2. *For all nonnegative integers N, and multi-indices α and β, there exists a constant A, depending on m, n, N, α and β only, such that*

$$\int_{\mathbb{R}^n} |z|^N \left| (\partial_x^\beta \partial_z^\alpha K_k)(x, z) \right| dz \le A2^{(m+|\alpha|-N)k}$$

for all $k = 0, 1, 2, \ldots$.

In the proof of Theorem 6.2, we make use of the following inequality:

$$|z|^{2N} \leq n^N \sum_{|\gamma|=N} |z^\gamma|^2 , \quad z \in \mathbb{R}^n . \tag{6.10}$$

The proof of (6.10) is left as an exercise. See Exercise 6.2.

Proof of Theorem 6.2. Let γ be any multi-index. Then, by (6.8), (6.9), Plancherel's theorem, Proposition 3.2, Leibnitz' formula, parts (iv) and (v) of Theorem 6.1,

$$\int_{\mathbb{R}^n} \left| z^\gamma \left(\partial_x^\beta \partial_z^\alpha K_k \right)(x,z) \right|^2 dz$$

$$= \int_{\mathbb{R}^n} \left| \partial_\xi^\gamma \left\{ \xi^\alpha \left(\partial_x^\beta \sigma_k \right)(x,\xi) \right\} \right|^2 d\xi$$

$$= \int_{W_k} \left| \sum_{\gamma' \leq \gamma} \binom{\gamma}{\gamma'} \partial_\xi^{\gamma'} \left\{ \xi^\alpha \left(\partial_x^\beta \sigma \right)(x,\xi) \right\} \left(\partial_\xi^{\gamma-\gamma'} \varphi_k \right)(\xi) \right|^2 d\xi ,$$
$$\tag{6.11}$$

where

$$W_0 = \{ \xi \in \mathbb{R}^n : |\xi| \leq 2 \} \tag{6.12}$$

and

$$W_k = \{ \xi \in \mathbb{R}^n : 2^{k-2} \leq |\xi| \leq 2^{k+1} \} \tag{6.13}$$

for $k = 1, 2, \ldots$. Hence, by (6.11), part (vi) of Theorem 6.1 and the fact that $\xi^\alpha (\partial_x^\beta \sigma)$ is a symbol in $S^{m+|\alpha|}$, we get positive constants $C_{\alpha,\beta,\gamma'}$ and $C_{\gamma,\gamma'}$ such that

$$\int_{\mathbb{R}^n} \left| z^\gamma \left(\partial_x^\beta \partial_z^\alpha K_k \right)(x,z) \right|^2 dz$$

$$\leq \int_{W_k} \left\{ \sum_{\gamma' \leq \gamma} \binom{\gamma}{\gamma'} C_{\alpha,\beta,\gamma'} (1 + |\xi|)^{m+|\alpha|-|\gamma'|} C_{\gamma,\gamma'} 2^{-k|\gamma-\gamma'|} \right\}^2 d\xi .$$
$$\tag{6.14}$$

Hence, by (6.12), (6.13) and (6.14),

$$\int_{\mathbb{R}^n} \left| z^\gamma \left(\partial_x^\beta \partial_z^\alpha K_k \right) (x, z) \right|^2 dz$$

$$\leq \int_{W_k} \left\{ \sum_{\gamma' \leq \gamma} \binom{\gamma}{\gamma'} C_{\alpha, \beta, \gamma'} 2^{(k+2)(m+|\alpha|-|\gamma'|)} C_{\gamma, \gamma'} 2^{-k|\gamma-\gamma'|} \right\}^2 d\xi$$

$$= 2^{k(2m+2|\alpha|-2|\gamma|)} \int_{W_k} \left\{ \sum_{\gamma' \leq \gamma} \binom{\gamma}{\gamma'} C_{\alpha, \beta, \gamma'} C_{\gamma, \gamma'} \right.$$
$$\left. 2^{2(m+|\alpha|-|\gamma'|)} \right\}^2 d\xi$$

$$\leq 2^{k(n+2m+2|\alpha|-2|\gamma|)} C_n 2^n \left\{ \sum_{\gamma' \leq \gamma} \binom{\gamma}{\gamma'} C_{\alpha, \beta, \gamma'} C_{\gamma, \gamma'} \right.$$
$$\left. 2^{2(m+|\alpha|-|\gamma'|)} \right\}^2, \tag{6.15}$$

where C_n is the constant with the property that the volume of the ball in $\mathbb{R}^n$ with radius r is equal to $C_n r^n$. Let $A_{\alpha, \beta, \gamma, m, n}$ be defined by

$$A_{\alpha, \beta, \gamma, m, n} = C_n 2^n \left\{ \sum_{\gamma' \leq \gamma} \binom{\gamma}{\gamma'} C_{\alpha, \beta, \gamma'} C_{\gamma, \gamma'} 2^{2(m+|\alpha|-|\gamma'|)} \right\}^2.$$

Then (6.15) becomes

$$\int_{\mathbb{R}^n} \left| z^\gamma \left(\partial_x^\beta \partial_z^\alpha K_k \right) (x, z) \right|^2 dz \leq A_{\alpha, \beta, \gamma, m, n} 2^{k(n+2m+2|\alpha|-2|\gamma|)}. \tag{6.16}$$

Let N be any nonnegative integer. Then, by (6.10) and (6.16),

$$\int_{\mathbb{R}^n} |z|^{2N} \left| \left(\partial_x^\beta \partial_z^\alpha K_k \right) (x, z) \right|^2 dz$$

$$\leq n^N \sum_{|\gamma|=N} \int_{\mathbb{R}^n} \left| z^\gamma \left(\partial_x^\beta \partial_z^\alpha K_k \right) (x, z) \right|^2 dz$$

$$\leq A_1 2^{k(n+2m+2|\alpha|-2N)}$$

for all $k = 0, 1, 2, \ldots$, where

$$A_1 = n^N \sum_{|\gamma| = N} A_{\alpha, \beta, \gamma, m, n} \; .$$

By taking square roots, we get

$$\left\{ \int_{\mathbb{R}^n} |z|^{2N} \left| \left(\partial_x^\beta \partial_z^\alpha K_k \right)(x, z) \right|^2 dz \right\}^{\frac{1}{2}} \leq A_2 2^{\left(\frac{n}{2} + m + |\alpha| - N \right) k} \quad (6.17)$$

for $k = 0, 1, 2, \ldots$, where $A_2 = A_1^{1/2}$. Now write

$$\int_{\mathbb{R}^n} |z|^N \left| \left(\partial_x^\beta \partial_z^\alpha K_k \right)(x, z) \right| dz = \int_{|z| \leq 2^{-k}} + \int_{|z| > 2^{-k}} \; . \quad (6.18)$$

By (6.17) and the Cauchy-Schwarz inequality, there exists a constant $A_3 > 0$, depending on m, n, N, α and β only, such that

$$\int_{|z| \leq 2^{-k}} \leq \left\{ \int_{\mathbb{R}^n} |z|^{2N} \left| \left(\partial_x^\beta \partial_z^\alpha K_k \right)(x, z) \right|^2 dz \right\}^{\frac{1}{2}} \left\{ \int_{|z| \leq 2^{-k}} dz \right\}^{\frac{1}{2}}$$

$$\leq A_3 2^{\left(\frac{n}{2} + m + |\alpha| - N \right) k} 2^{-\frac{nk}{2}}$$

$$= A_3 2^{(m + |\alpha| - N) k} \quad (6.19)$$

for $k = 0, 1, 2, \ldots$. Next, by (6.17) and the Cauchy-Schwarz inequality again, there is another constant $A_4 > 0$, depending on m, n, N, α and β only, such that

$$\int_{|z| > 2^{-k}}$$

$$\leq \left\{ \int_{\mathbb{R}^n} |z|^{2(N+n)} \left| \left(\partial_x^\beta \partial_z^\alpha K_k \right)(x, z) \right|^2 dz \right\}^{\frac{1}{2}} \left\{ \int_{|z| > 2^{-k}} |z|^{-2n} dz \right\}^{\frac{1}{2}}$$

$$\leq A_4 2^{\left(\frac{n}{2} + m + |\alpha| - N - n \right) k} \left\{ \int_{S^{n-1}} \int_{2^{-k}}^\infty r^{-2n} r^{n-1} dr d\sigma \right\}^{\frac{1}{2}} \; ,$$

where $d\sigma$ is the surface measure on the unit sphere S^{n-1}. Hence

$$\int_{|z|>2^{-k}} \leq A_4 2^{(\frac{n}{2}+m+|\alpha|-N-n)k} \left|S^{n-1}\right|^{\frac{1}{2}} n^{-\frac{1}{2}} 2^{\frac{nk}{2}}$$

$$= A_5 2^{(m+|\alpha|-N)k} \tag{6.20}$$

for $k = 0, 1, 2, \ldots$, where $A_5 = A_4 |S^{n-1}|^{\frac{1}{2}} n^{-\frac{1}{2}}$ and $|S^{n-1}|$ is the surface area of S^{n-1}. Hence, by (6.18), (6.19) and (6.20), we complete the proof of Theorem 6.2.

The following version of Taylor's formula with integral remainder plays an important role in the study of pseudo-differential operators.

Theorem 6.3.   *Let $f \in C^\infty(\mathbb{R}^n)$. Then for all positive integers N,*

$$f(\xi + \eta)$$
$$= \sum_{|\alpha|<N} \frac{(\partial^\alpha f)(\xi)}{\alpha!} \eta^\alpha + N \sum_{|\gamma|=N} \frac{\eta^\gamma}{\gamma!} \int_0^1 (1-\theta)^{N-1} (\partial^\gamma f)(\xi + \theta\eta)\, d\theta$$

$$\tag{6.21}$$

for all $\xi, \eta \in \mathbb{R}^n$.

Proof: The proof is by induction on N. For $N = 1$, we need to prove that

$$f(\xi + \eta) = f(\xi) + \sum_{|\gamma|=1} \frac{\eta^\gamma}{\gamma!} \int_0^1 (\partial^\gamma f)(\xi + \theta\eta)\, d\theta \tag{6.22}$$

for all $\xi, \eta \in \mathbb{R}^n$. To do this, we define a function $\varphi : \mathbb{R} \to \mathbb{C}$ by

$$\varphi(x) = f(\xi + x\eta), \quad x \in \mathbb{R}. \tag{6.23}$$

Then, by the fundamental theorem of calculus,

$$\varphi(1) = \varphi(0) + \int_0^1 \varphi'(t)\, dt. \tag{6.24}$$

Hence, by (6.23), (6.24) and the chain rule, we get the formula

$$f(\xi + \eta) = f(\xi) + \sum_{j=1}^{n} \int_0^1 (\partial_j f)(\xi + t\eta)\eta_j dt , \quad \xi, \eta \in \mathbb{R}^n ,$$

which is exactly the same as (6.22). We now assume that (6.21) is true for the positive integer N. Then, by the induction hypothesis, we have

$$f(\xi + \eta) - \sum_{|\alpha| < N+1} \frac{(\partial^\alpha f)(\xi)}{\alpha!}\eta^\alpha$$

$$= f(\xi + \eta) - \sum_{|\alpha| < N} \frac{(\partial^\alpha f)(\xi)}{\alpha!}\eta^\alpha - \sum_{|\alpha| = N} \frac{(\partial^\alpha f)(\xi)}{\alpha!}\eta^\alpha$$

$$= N \sum_{|\gamma| = N} \frac{\eta^\gamma}{\gamma!} \int_0^1 (1 - \theta)^{N-1} (\partial^\gamma f)(\xi + \theta\eta) \, d\theta$$

$$- \sum_{|\alpha| = N} \frac{(\partial^\alpha f)(\xi)}{\alpha!}\eta^\alpha \tag{6.25}$$

for all $\xi, \eta \in \mathbb{R}^n$. Now, using the fact that

$$N \int_0^1 (1 - \theta)^{N-1} d\theta = 1$$

and the formula (6.21) for $N = 1$, the formula (6.25) becomes

$$f(\xi + \eta) - \sum_{|\alpha| < N+1} \frac{(\partial^\alpha f)(\xi)}{\alpha!}\eta^\alpha$$

$$= \sum_{|\gamma| = N} \frac{\eta^\gamma}{\gamma!} N \int_0^1 (1 - \theta)^{N-1} \{(\partial^\gamma f)(\xi + \theta\eta) - (\partial^\gamma f)(\xi)\} \, d\theta$$

$$= \sum_{|\gamma| = N} \frac{\eta^\gamma}{\gamma!} N \int_0^1 (1 - \theta)^{N-1} \sum_{|\delta| = 1} \frac{(\theta\eta)^\delta}{\delta!}$$

$$\left\{ \int_0^1 (\partial^{\gamma+\delta} f)(\xi + \varphi\theta\eta) d\varphi \right\} d\theta . \tag{6.26}$$

Let $\rho = \varphi\theta$ in (6.26). Then

$$
f(\xi + \eta) - \sum_{|\alpha| < N+1} \frac{(\partial^\alpha f)(\xi)}{\alpha!} \eta^\alpha
$$
$$
= \sum_{|\gamma|=N} \sum_{|\delta|=1} \frac{\eta^{\gamma+\delta}}{\gamma!\delta!} N \int_0^1 (1-\theta)^{N-1} \left\{ \int_0^\theta (\partial^{\gamma+\delta} f)(\xi + \rho\eta) d\rho \right\} d\theta .
$$
$$
(6.27)
$$

By interchanging the order of integration in (6.27), we have

$$
f(\xi + \eta) - \sum_{|\alpha| < N+1} \frac{(\partial^\alpha f)(\xi)}{\alpha!} \eta^\alpha
$$
$$
= \sum_{|\gamma|=N} \sum_{|\delta|=1} \frac{\eta^{\gamma+\delta}}{\gamma!\delta!} N \int_0^1 (\partial^{\gamma+\delta} f)(\xi + \rho\eta) \left\{ \int_\rho^1 (1-\theta)^{N-1} d\theta \right\} d\rho
$$
$$
= \sum_{|\gamma|=N} \sum_{|\delta|=1} \frac{\eta^{\gamma+\delta}}{\gamma!\delta!} \int_0^1 (\partial^{\gamma+\delta} f)(\xi + \rho\eta)(1-\rho)^N d\rho . \qquad (6.28)
$$

For all multi-indices γ and δ with $|\gamma| = N$ and $|\delta| = 1$, we have

$$
\frac{(\gamma + \delta)!}{\gamma!\delta!} = \gamma \cdot \delta + 1 , \qquad (6.29)
$$

where $\gamma \cdot \delta$ is the inner product of γ and δ. Hence, by (6.28) and (6.29),

$$
f(\xi + \eta) - \sum_{|\alpha| < N+1} \frac{(\partial^\alpha f)(\xi)}{\alpha!} \eta^\alpha
$$
$$
= \sum_{|\gamma|=N} \sum_{|\delta|=1} \frac{\eta^{\gamma+\delta}}{(\gamma + \delta)!} (\gamma \cdot \delta + 1) \int_0^1 (\partial^{\gamma+\delta} f)(\xi + \rho\eta)(1-\rho)^N d\rho .
$$
$$
(6.30)
$$

For each multi-index α of length $N + 1$, we can write α in the form $\gamma + \delta$, where $|\gamma| = N$ and $|\delta| = 1$ by picking

$$
\gamma = (\alpha_1, \alpha_2, \ldots, \alpha_j - 1, \ldots, \alpha_n) \qquad (6.31)
$$

and

$$\delta = (0, 0, \ldots, 1, \ldots, 0) \tag{6.32}$$

in which δ has 1 in the jth position whenever $\alpha_j \geq 1$. Then, by (6.31) and (6.32), we have

$$\gamma \cdot \delta = \alpha_j - 1 \tag{6.33}$$

whenever $\alpha_j \geq 1$. Hence, by (6.30) and (6.33),

$$f(\xi + \eta) - \sum_{|\alpha| < N+1} \frac{(\partial^\alpha f)(\xi)}{\alpha!} \eta^\alpha$$

$$= \sum_{|\alpha| = N+1} \sum_{\alpha_j \geq 1} \frac{\eta^\alpha}{\alpha!} \alpha_j \int_0^1 (\partial^\alpha f)(\xi + \rho\eta)(1 - \rho)^N d\rho$$

$$= (N + 1) \sum_{|\alpha| = N+1} \frac{\eta^\alpha}{\alpha!} \int_0^1 (\partial^\alpha f)(\xi + \rho\eta)(1 - \rho)^N d\rho \, .$$

Therefore (6.21) is also true for the integer $N + 1$. Hence, by the principle of mathematical induction, the proof of Theorem 6.3 is complete.

Exercises

6.1. Construct a function ψ in $C_0^\infty(\mathbb{R}^n)$ such that $0 \leq \psi(\xi) \leq 1$ for $\xi \in \mathbb{R}^n$, $\psi(\xi) = 1$ for $1 \leq |\xi| \leq 2$ and $\psi(\xi) = 0$ for $|\xi| < \frac{1}{2}$ or $|\xi| > 4$.

6.2. For any nonnegative integer N, prove that

$$|z|^{2N} \leq n^N \sum_{|\gamma| = N} |z^\gamma|^2$$

for all $z \in \mathbb{R}^n$.

7. THE PRODUCT OF TWO PSEUDO-DIFFERENTIAL OPERATORS

In this chapter we prove that the product (or composition) of two pseudo-differential operators is again a pseudo-differential operator. We also give an asymptotic expansion for the symbol of the product. The main result is the following theorem:

Theorem 7.1. *Let $\sigma \in S^{m_1}$ and $\tau \in S^{m_2}$. Then the product $T_\sigma T_\tau$ of the pseudo-differential operators T_σ and T_τ is again a pseudo-differential operator T_λ, where λ is in $S^{m_1+m_2}$ and has the following asymptotic expansion:*

$$\lambda \sim \sum_\mu \frac{(-i)^{|\mu|}}{\mu!} \left(\partial_\xi^\mu \sigma \right) \left(\partial_x^\mu \tau \right) . \tag{7.1}$$

Here (7.1) means that

$$\lambda - \sum_{|\mu|<N} \frac{(-i)^{|\mu|}}{\mu!} \left(\partial_\xi^\mu \sigma \right) \left(\partial_x^\mu \tau \right)$$

is a symbol in $S^{m_1+m_2-N}$ for every positive integer N.

63

To motivate the proof, let us begin with an argument to find out how we should proceed. For any function φ in $\mathcal{S}$, we have

$$(T_{\sigma_k}\varphi)(x) = (2\pi)^{-n/2}\int_{\mathbb{R}^n} e^{ix\cdot\xi}\sigma_k(x,\xi)\hat{\varphi}(\xi)d\xi$$

for all $x \in \mathbb{R}^n$, where $\sigma_k(x,\xi) = \sigma(x,\xi)\varphi_k(\xi)$ and $\{\varphi_k\}$ is the partition of unity constructed in Theorem 6.1. Hence

$$\sum_{k=0}^{\infty}(T_{\sigma_k}\varphi)(x) = (2\pi)^{-n/2}\int_{\mathbb{R}^n} e^{ix\cdot\xi}\left\{\sum_{k=0}^{\infty}\sigma_k(x,\xi)\right\}\hat{\varphi}(\xi)d\xi$$

$$= (2\pi)^{-n/2}\int_{\mathbb{R}^n} e^{ix\cdot\xi}\sigma(x,\xi)\hat{\varphi}(\xi)d\xi$$

for all $x \in \mathbb{R}^n$. The interchange of $\sum_{k=0}^{\infty}$ and $\int_{\mathbb{R}^n}$ can of course be justified by Fubini's theorem. Hence

$$T_\sigma\varphi = \sum_{k=0}^{\infty}T_{\sigma_k}\varphi$$

and the convergence of the series can be shown to be absolute and uniform for all $x \in \mathbb{R}^n$. Our goal is to compute the symbol of $T_\sigma T_\tau$. But let us first compute the symbol of $T_{\sigma_k}T_\tau$. You will see why in a minute! Let $\varphi \in \mathcal{S}$. Then, by the definitions of a pseudo-differential operator and the Fourier transform, Fubini's theorem and (6.9), we get

$$(T_{\sigma_k}T_\tau\varphi)(x)$$

$$= (2\pi)^{-n/2}\int_{\mathbb{R}^n} e^{ix\cdot\xi}\sigma_k(x,\xi)(T_\tau\varphi)\hat{}(\xi)d\xi$$

$$= (2\pi)^{-n}\int_{\mathbb{R}^n} e^{ix\cdot\xi}\sigma_k(x,\xi)\left\{\int_{\mathbb{R}^n} e^{-i\xi\cdot y}(T_\tau\varphi)(y)dy\right\}d\xi$$

$$= (2\pi)^{-n}\int_{\mathbb{R}^n}\left\{\int_{\mathbb{R}^n} e^{i(x-y)\cdot\xi}\sigma_k(x,\xi)d\xi\right\}(T_\tau\varphi)(y)dy$$

$$= (2\pi)^{-n/2}\int_{\mathbb{R}^n} K_k(x, x-y)(T_\tau\varphi)(y)dy \qquad (7.2)$$

for all $x \in \mathbb{R}^n$. Hence, by the definition of a pseudo-differential operator and Fubini's theorem again, the formula (7.2) becomes

$$
(T_{\sigma_k} T_\tau \varphi)(x)
$$
$$
= (2\pi)^{-n} \int_{\mathbb{R}^n} K_k(x, x - y) \left\{ \int_{\mathbb{R}^n} e^{iy \cdot \eta} \tau(y, \eta) \hat{\varphi}(\eta) d\eta \right\} dy
$$
$$
= (2\pi)^{-n} \int_{\mathbb{R}^n} e^{ix \cdot \eta} \left\{ \int_{\mathbb{R}^n} e^{-i(x-y) \cdot \eta} K_k(x, x - y) \tau(y, \eta) dy \right\} \hat{\varphi}(\eta) d\eta
$$
$$
= (2\pi)^{-n/2} \int_{\mathbb{R}^n} e^{ix \cdot \eta} \lambda_k(x, \eta) \hat{\varphi}(\eta) d\eta
$$

for all $x \in \mathbb{R}^n$, where

$$
\lambda_k(x, \eta) = (2\pi)^{-n/2} \int_{\mathbb{R}^n} e^{-i(x-y) \cdot \eta} K_k(x, x - y) \tau(y, \eta) dy .
$$

By a simple change of variable,

$$
\lambda_k(x, \eta) = (2\pi)^{-n/2} \int_{\mathbb{R}^n} e^{-iz \cdot \eta} K_k(x, z) \tau(x - z, \eta) dz \qquad (7.3)
$$

for all $x, \eta \in \mathbb{R}^n$. This suggests that

$$
(T_\sigma T_\tau \varphi)(x) = (2\pi)^{-n/2} \int_{\mathbb{R}^n} e^{ix \cdot \eta} \lambda(x, \eta) \hat{\varphi}(\eta) d\eta
$$

for all $x \in \mathbb{R}^n$, where

$$
\lambda(x, \eta) = \sum_{k=0}^{\infty} \lambda_k(x, \eta) \qquad (7.4)
$$

for all $x, \eta \in \mathbb{R}^n$. Hence all we need to do is to show that $\lambda(x, \eta)$, defined by (7.3) and (7.4), is a symbol in $S^{m_1 + m_2}$ and satisfies (7.1).

Remark 7.2. Had we begun with $(T_\sigma T_\tau \varphi)(x)$ instead of $(T_{\sigma_k} T_\tau \varphi)(x)$, we would have the divergent integral $\int_{\mathbb{R}^n} e^{i(x-y) \cdot \xi} \sigma(x, \xi) d\xi$ instead of $\int_{\mathbb{R}^n} e^{i(x-y) \cdot \xi} \sigma_k(x, \xi) d\xi$ in (7.2).

This is why we cut off the symbol $\sigma(x,\xi)$ in the ξ variable by the partition of unity $\{\varphi_k\}$.

Proof of Theorem 7.1.   For $k = 0, 1, 2, \ldots$, we define λ_k by

$$\lambda_k(x,\xi) = (2\pi)^{-n/2} \int_{\mathbb{R}^n} e^{-iz\cdot\xi} K_k(x,z)\tau(x-z,\xi)dz \qquad (7.5)$$

for all $x,\xi \in \mathbb{R}^n$. Now, by the Taylor's formula with integral remainder given in Theorem 6.3, we get

$$\tau(x-z,\xi) = \sum_{|\mu|<N_1} \frac{(-z)^\mu}{\mu!} \left(\partial_x^\mu\tau\right)(x,\xi) + R_{N_1}(x,z,\xi) , \qquad (7.6)$$

where

$$R_{N_1}(x,z,\xi) = N_1 \sum_{|\mu|=N_1} \frac{(-z)^\mu}{\mu!} \int_0^1 (1-\theta)^{N_1-1} \left(\partial_x^\mu\tau\right)(x-\theta z,\xi)d\theta$$

$$(7.7)$$

for all $x,z,\xi \in \mathbb{R}^n$. Replacing $\tau(x-z,\xi)$ in (7.5) by the right hand side of (7.6), and using (6.8), (6.9), Proposition 3.2 and the Fourier inversion formula, we get

$$\lambda_k(x,\xi) = \sum_{|\mu|<N_1} \frac{(-i)^{|\mu|}}{\mu!} \left(\partial_\xi^\mu\sigma_k\right)(x,\xi)(\partial_x^\mu\tau)(x,\xi) + T_{N_1}^{(k)}(x,\xi) ,$$

$$(7.8)$$

where

$$T_{N_1}^{(k)}(x,\xi) = (2\pi)^{-n/2} \int_{\mathbb{R}^n} e^{-iz\cdot\xi} K_k(x,z) R_{N_1}(x,z,\xi)dz \qquad (7.9)$$

for all $x,\xi \in \mathbb{R}^n$. For any positive integer N, the function λ given by (7.4) satisfies

$$\lambda - \sum_{|\mu|<N} \frac{(-i)^{|\mu|}}{\mu!} \left(\partial_\xi^\mu\sigma\right)(\partial_x^\mu\tau)$$

$$= \lambda - \sum_{|\mu|<N_1} \frac{(-i)^{|\mu|}}{\mu!} \left(\partial_\xi^\mu\sigma\right)(\partial_x^\mu\tau)$$

$$+ \sum_{N\leq|\mu|<N_1} \frac{(-i)^{|\mu|}}{\mu!} \left(\partial_\xi^\mu\sigma\right)(\partial_x^\mu\tau) , \qquad (7.10)$$

where N_1 is any integer larger than N. Obviously,

$$\sum_{N \leq |\mu| < N_1} \frac{(-i)^{|\mu|}}{\mu!} \left(\partial_\xi^\mu \sigma \right) \left(\partial_x^\mu \tau \right) \in S^{m_1 + m_2 - N} \ .$$

Hence, if we can prove that for all multi-indices α and β, there exists a constant $C_{\alpha, \beta} > 0$ such that

$$\left| \left\{ D_x^\alpha D_\xi^\beta \left[\lambda - \sum_{|\mu| < N_1} \frac{(-i)^{|\mu|}}{\mu!} \left(\partial_\xi^\mu \sigma \right) \left(\partial_x^\mu \tau \right) \right] \right\} (x, \xi) \right|$$
$$\leq C_{\alpha, \beta} (1 + |\xi|)^{m_1 + m_2 - N - |\beta|} \tag{7.11}$$

for all $x, \xi \in \mathbb{R}^n$, then we can conclude that λ is in $S^{m_1 + m_2}$ and has an asymptotic expansion given by (7.1). To this end, we first make note that, by (6.8), (7.4) and (7.8),

$$\lambda - \sum_{|\mu| < N_1} \frac{(-i)^{|\mu|}}{\mu!} \left(\partial_\xi^\mu \sigma \right) \left(\partial_x^\mu \tau \right) = \sum_{k=0}^\infty T_{N_1}^{(k)} \ . \tag{7.12}$$

Then, for any two multi-indices α and β, we need to estimate $D_x^\alpha D_\xi^\beta T_{N_1}^{(k)}$ for all $k = 0, 1, 2, \ldots$. We have the following estimate:

Lemma 7.3. *For all nonnegative integers M, there exists a positive constant $C_{\alpha, \beta, M, N_1}$ such that*

$$\left| \left(D_x^\alpha D_\xi^\beta T_{N_1}^{(k)} \right) (x, \xi) \right|$$
$$\leq C_{\alpha, \beta, M, N_1} (1 + |\xi|)^{m_2 - 2M} 2^{(m_1 + 2M - N_1)k} \tag{7.13}$$

for all $x, \xi \in \mathbb{R}^n$ and $k = 0, 1, 2, \ldots$.

Let us assume Lemma 7.3 for a moment. Then, for all positive integers N, and multi-indices α and β, we can choose a positive integer M such that

$$(1 + |\xi|)^{m_2 - 2M} \leq (1 + |\xi|)^{m_1 + m_2 - N - |\beta|} \tag{7.14}$$

for all $\xi \in \mathbb{R}^n$. With this M fixed, we can choose another positive integer N_1 so large that

$$m_1 + 2M - N_1 < 0 . \tag{7.15}$$

By (7.12)–(7.15),

$$\left| \left\{ D_x^\alpha D_\xi^\beta \left(\lambda - \sum_{|\mu| < N_1} \frac{(-i)^{|\mu|}}{\mu!} \left(\partial_\xi^\mu \sigma \right) \left(\partial_x^\mu \tau \right) \right) \right\} (x, \xi) \right|$$

$$\leq \sum_{k=0}^\infty C_{\alpha,\beta,M,N_1} (1 + |\xi|)^{m_1 + m_2 - N - |\beta|} 2^{(m_1 + 2M - N_1)k}$$

$$= C_{\alpha,\beta} (1 + |\xi|)^{m_1 + m_2 - N - |\beta|}$$

for all $x, \xi \in \mathbb{R}^n$, where

$$C_{\alpha,\beta} = C_{\alpha,\beta,M,N_1} \sum_{k=0}^\infty 2^{(m_1 + 2M - N_1)k} .$$

It remains to prove Lemma 7.3. To this end, we need another lemma.

Lemma 7.4.   *Let $R_{N_1}(x, z, \xi)$ be the function given by (7.7). Then, for all multi-indices α, β and γ, there exists a constant $C_{\alpha,\beta,\gamma} > 0$ such that*

$$\left| \left(\partial_z^\gamma \partial_x^\alpha \partial_\xi^\beta R_{N_1} \right) (x, z, \xi) \right| \leq C_{\alpha,\beta,\gamma} \left[\sum_{\gamma' \leq \gamma} |z|^{N_1 - |\gamma'|} \right] (1 + |\xi|)^{m_2 - |\beta|}$$

for all $x, z, \xi \in \mathbb{R}^n$.

The proof that Lemma 7.4 implies Lemma 7.3 is by Leibnitz' formula, integration by parts, Exercise 3.5 and Theorem 6.2. We leave it as an exercise. See Exercise 7.1.

Proof of Lemma 7.4. By (7.7), we have

$$
\left(\partial_x^\alpha \partial_\xi^\beta R_{N_1} \right)(x, z, \xi)
$$

$$
= N_1 \sum_{|\mu|=N_1} \frac{(-z)^\mu}{\mu!} \int_0^1 (1-\theta)^{N_1-1} \left(\partial_x^{\alpha+\mu} \partial_\xi^\beta \tau \right)(x - \theta z, \xi) d\theta
$$

(7.16)

for all $x, z, \xi \in \mathbb{R}^n$. Hence, by (7.16) and Leibnitz' formula,

$$
\left(\partial_z^\gamma \partial_x^\alpha \partial_\xi^\beta R_{N_1} \right)(x, z, \xi)
$$

$$
= N_1 \sum_{|\mu|=N_1} \sum_{\gamma' \leq \gamma} \binom{\gamma}{\gamma'} \frac{1}{\mu!} \left\{ \partial_z^{\gamma'} (-z)^\mu \right\} \int_0^1 (1-\theta)^{N_1-1}
$$

$$
\left(\partial_x^{\gamma-\gamma'+\alpha+\mu} \partial_\xi^\beta \tau \right)(x - \theta z, \xi)(-\theta)^{|\gamma-\gamma'|} d\theta
$$

(7.17)

for all $x, z, \xi \in \mathbb{R}^n$. So, by Exercise 3.5, (7.17) and the fact that $\tau \in S^{m_2}$, there exist positive constants $C_{\gamma'}$ and $C_{\alpha,\beta,\gamma',\mu}$ such that

$$
\left| \left(\partial_z^\gamma \partial_x^\alpha \partial_\xi^\beta R_{N_1} \right)(x, z, \xi) \right|
$$

$$
\leq N_1 \sum_{|\mu|=N_1} \sum_{\gamma' \leq \gamma} \binom{\gamma}{\gamma'} C_{\gamma'} |z|^{N_1-|\gamma'|} \int_0^1 C_{\alpha,\beta,\gamma',\mu} (1+|\xi|)^{m_2-|\beta|} d\theta
$$

$$
\leq C_{\alpha,\beta,\gamma} \left[\sum_{\gamma' \leq \gamma} |z|^{N_1-|\gamma'|} \right] (1+|\xi|)^{m_2-|\beta|}
$$

for all $x, z, \xi \in \mathbb{R}^n$, where

$$
C_{\alpha,\beta,\gamma} = N_1 \sum_{|\mu|=N_1} \sup_{\gamma' \leq \gamma} \left\{ \binom{\gamma}{\gamma'} C_{\gamma'} C_{\alpha,\beta,\gamma',\mu} \right\}
$$

and this completes the proof of Lemma 7.4.

Exercises

7.1.   Prove that Lemma 7.4 implies Lemma 7.3.

7.2.   Let $P(x, D) = \sum\limits_{|\alpha| \leq m_1} a_\alpha(x) D^\alpha$ and $Q(x, D) = \sum\limits_{|\alpha| \leq m_2} b_\alpha(x) D^\alpha$, where the a_α's and b_α's are in $C^\infty(\mathbb{R}^n)$ and all their partial derivatives are bounded functions on $\mathbb{R}^n$. Compute the symbol of the product $P(x, D)Q(x, D)$ directly. Compare the answer with the symbol obtained by Theorem 7.1.

7.3.   Let $q \in C^\infty(\mathbb{R}^n)$ be such that

$$\sup_{x \in \mathbb{R}^n} |(D^\alpha q)(x)| < \infty$$

for all multi-indices α. Let σ be the symbol defined by

$$\sigma(x, \xi) = q(x)$$

for all $x, \xi \in \mathbb{R}^n$. Let τ be any other symbol. Use Theorem 7.1 to compute the symbols of the operators $T_\sigma T_\tau$ and $T_\tau T_\sigma$.

7.4.   Let $\sigma \in S^{m_1}$ and $\tau \in S^{m_2}$. Prove that the symbol of the pseudo-differential operator $T_\sigma T_\tau - T_\tau T_\sigma$ is in $S^{m_1 + m_2 - 1}$.

8. THE FORMAL ADJOINT OF A PSEUDO-DIFFERENTIAL OPERATOR

We begin with a notation. For any pair of functions φ and ψ in $\mathcal{S}$, we define (φ, ψ) by

$$(\varphi, \psi) = \int_{\mathbb{R}^n} \varphi(x)\overline{\psi(x)}dx . \tag{8.1}$$

Let σ be a symbol in S^m and T_σ its associated pseudo-differential operator. Suppose there exists a linear operator T_σ^* : $\mathcal{S} \to \mathcal{S}$ such that

$$(T_\sigma \varphi, \psi) = (\varphi, T_\sigma^* \psi) , \quad \varphi, \psi \in \mathcal{S} . \tag{8.2}$$

Then we call T_σ^* a *formal adjoint* of the operator T_σ. It is very easy to see that a pseudo-differential operator has at most one formal adjoint. Three problems arise.

1. Does a formal adjoint exist?
2. If it exists, is it a pseudo-differential operator?
3. If it is a pseudo-differential operator, can we find an asymptotic expansion for its symbol?

The aim of this chapter is to prove that the formal adjoint of a pseudo-differential operator exists and is a pseudo-differential

71

operator. Moreover, we can obtain a useful asymptotic expansion for the symbol of the formal adjoint. To be more precise, let us prove the following theorem:

Theorem 8.1.   *Let σ be a symbol in S^m. Then the formal adjoint of the pseudo-differential operator T_σ is again a pseudo-differential operator T_τ, where τ is a symbol in S^m and has the following asymptotic expansion:*

$$\tau(x,\xi) \sim \sum_\mu \frac{(-i)^{|\mu|}}{\mu!} \left(\partial_x^\mu \partial_\xi^\mu \bar{\sigma} \right)(x,\xi) . \tag{8.3}$$

Here (8.3) means that

$$\tau(x,\xi) - \sum_{|\mu|<N} \frac{(-i)^{|\mu|}}{\mu!} \left(\partial_x^\mu \partial_\xi^\mu \bar{\sigma} \right)(x,\xi)$$

is a symbol in S^{m-N} for every positive integer N.

Before the proof, let us show how the symbol τ can be constructed. For $k = 0, 1, 2, \ldots$, we define σ_k and K_k by (6.8) and (6.9) respectively. Then, by the definition of a formal adjoint,

$$(T_{\sigma_k}\varphi, \psi) = \left(\varphi, T_{\sigma_k}^* \psi\right) \tag{8.4}$$

for all φ and ψ in $\mathcal{S}$. By (8.1), Proposition 3.4 (ii), Proposition 3.6 and the definition of a pseudo-differential operator,

$$(T_{\sigma_k}\varphi, \psi) = \int_{\mathbb{R}^n} (T_{\sigma_k}\varphi)(x)\overline{\psi(x)}dx$$

$$= (2\pi)^{-n/2} \int_{\mathbb{R}^n} \left\{ \int_{\mathbb{R}^n} e^{ix\cdot\eta} \sigma_k(x,\eta)\hat{\varphi}(\eta)d\eta \right\} \overline{\psi(x)}dx$$

$$= (2\pi)^{-n/2} \int_{\mathbb{R}^n} \left\{ \int_{\mathbb{R}^n} \hat{\sigma}_k(x, y-x)\varphi(y)dy \right\} \overline{\psi(x)}dx , \tag{8.5}$$

where

$$\hat{\sigma}_k(x, y) = (2\pi)^{-n/2} \int_{\mathbb{R}^n} e^{-iy\cdot\eta} \sigma_k(x, \eta) d\eta \tag{8.6}$$

for all $x, y \in \mathbb{R}^n$. Therefore, by (6.9), (8.5), (8.6) and Fubini's theorem,

$$
\begin{aligned}
(T_{\sigma_k}\varphi, \psi) &= (2\pi)^{-n/2} \int_{\mathbb{R}^n} \left\{ \int_{\mathbb{R}^n} \hat{\sigma}_k(x, y - x)\overline{\psi(x)} dx \right\} \varphi(y) dy \\
&= (2\pi)^{-n/2} \int_{\mathbb{R}^n} \left\{ \int_{\mathbb{R}^n} K_k(x, x - y)\overline{\psi(x)} dx \right\} \varphi(y) dy
\end{aligned}
\tag{8.7}
$$

for all $\varphi, \psi \in \mathcal{S}$. Therefore, by (8.4) and (8.7),

$$\left(T_{\sigma_k}^* \psi\right)(x) = (2\pi)^{-n/2} \int_{\mathbb{R}^n} \overline{K_k}(y, y - x)\psi(y) dy \tag{8.8}$$

for all $x \in \mathbb{R}^n$. Hence, applying the Fourier inversion formula to the function ψ on the right hand side of (8.8), Fubini's theorem and a change of variables, we have

$$
\begin{aligned}
&\left(T_{\sigma_k}^* \psi\right)(x) \\
&= (2\pi)^{-n} \int_{\mathbb{R}^n} \overline{K_k}(y, y - x) \left\{ \int_{\mathbb{R}^n} e^{iy\cdot\eta} \hat{\psi}(\eta) d\eta \right\} dy \\
&= (2\pi)^{-n} \int_{\mathbb{R}^n} \left\{ \int_{\mathbb{R}^n} e^{iy\cdot\eta} \overline{K_k}(y, y - x) dy \right\} \hat{\psi}(\eta) d\eta \\
&= (2\pi)^{-n} \int_{\mathbb{R}^n} e^{ix\cdot\eta} \left\{ \int_{\mathbb{R}^n} e^{i\eta\cdot(y-x)} \overline{K_k}(y, y - x) dy \right\} \hat{\psi}(\eta) d\eta \\
&= (2\pi)^{-n} \int_{\mathbb{R}^n} e^{ix\cdot\eta} \left\{ \int_{\mathbb{R}^n} e^{i\eta\cdot z} \overline{K_k}(x + z, z) dz \right\} \hat{\psi}(\eta) d\eta
\end{aligned}
\tag{8.9}
$$

for all $x \in \mathbb{R}^n$. It is clear from (8.9) that

$$T_{\sigma_k}^* = T_{\tau_k}, \tag{8.10}$$

where

$$\tau_k(x, \eta) = (2\pi)^{-n/2} \int_{\mathbb{R}^n} e^{i\eta\cdot z} \overline{K_k}(x + z, z) dz. \tag{8.11}$$

Since $(T_\sigma \varphi, \psi) = \sum_{k=0}^{\infty} (T_{\sigma_k} \varphi, \psi)$ for all $\varphi, \psi \in \mathcal{S}$, it is clear from (8.4) and (8.10) that a good candidate for τ is given by

$$\tau(x, \eta) = \sum_{k=0}^{\infty} \tau_k(x, \eta) \tag{8.12}$$

for all $x, \eta \in \mathbb{R}^n$. Hence it remains to prove that τ is a symbol of order m with an asymptotic expansion given by (8.3) and

$$(T_\sigma \varphi, \psi) = (\varphi, T_\tau \psi) \tag{8.13}$$

for all $\varphi, \psi \in \mathcal{S}$.

Proof of Theorem 8.1.   For $k = 0, 1, 2, \ldots$, define τ_k by (8.11). Let N_1 be any positive integer. Then, by the Taylor's formula with integral remainder given in Theorem 6.3,

$$K_k(x + z, z) = \sum_{|\mu| < N_1} \frac{z^\mu}{\mu!} \left(\partial_x^\mu K_k\right)(x, z) + R_{N_1}^{(k)}(x, z) , \tag{8.14}$$

where, by (6.9),

$$R_{N_1}^{(k)}(x, z)$$

$$= N_1 \sum_{|\mu| = N_1} \frac{z^\mu}{\mu!} \int_0^1 (1 - \theta)^{N_1 - 1} \left(\partial_x^\mu K_k\right)(x + \theta z, z) d\theta$$

$$= N_1 \sum_{|\mu| = N_1} \frac{z^\mu}{\mu!} \int_0^1 (1 - \theta)^{N_1 - 1} (2\pi)^{-n/2}$$

$$\int_{\mathbb{R}^n} e^{iz \cdot \xi} \left(\partial_x^\mu \sigma_k\right)(x + \theta z, \xi) d\xi d\theta . \tag{8.15}$$

Hence, by (6.9), (8.11), (8.14) and Proposition 3.2,

$$\tau_k(x, \eta) = \sum_{|\mu| < N_1} \frac{(-i)^{|\mu|}}{\mu!} \left(\partial_x^\mu \partial_\eta^\mu \bar{\sigma}_k\right)(x, \eta) + T_{N_1}^{(k)}(x, \eta) , \tag{8.16}$$

where

$$T_{N_1}^{(k)}(x,\eta) = (2\pi)^{-n/2} \int_{\mathbb{R}^n} e^{i\eta \cdot z} \overline{R_{N_1}^{(k)}}(x,z)dz \qquad (8.17)$$

for all $x,\eta \in \mathbb{R}^n$. For any positive integer N,

$$\tau - \sum_{|\mu|<N} \frac{(-i)^{|\mu|}}{\mu!} \partial_x^\mu \partial_\eta^\mu \bar{\sigma}$$

$$= \tau - \sum_{|\mu|<N_1} \frac{(-i)^{|\mu|}}{\mu!} \partial_x^\mu \partial_\eta^\mu \bar{\sigma} + \sum_{N \leq |\mu|<N_1} \frac{(-i)^{|\mu|}}{\mu!} \partial_x^\mu \partial_\eta^\mu \bar{\sigma} , \qquad (8.18)$$

where N_1 is any integer larger than N. Obviously, $\sum_{N \leq |\mu|<N_1} \frac{(-i)^{|\mu|}}{\mu!}$ $\partial_x^\mu \partial_\eta^\mu \bar{\sigma}$ is in S^{m-N}. Hence, if we can prove that for all multi-indices α and β, there exists a constant $C_{\alpha,\beta} > 0$ such that

$$\left| \left[D_x^\alpha D_\eta^\beta \left\{ \tau - \sum_{|\mu|<N_1} \frac{(-i)^{|\mu|}}{\mu!} \partial_x^\mu \partial_\eta^\mu \bar{\sigma} \right\} \right](x,\eta) \right|$$

$$\leq C_{\alpha,\beta}(1+|\eta|)^{m-N-|\beta|} \qquad (8.19)$$

for all $x,\eta \in \mathbb{R}^n$, then we can conclude that $\tau \in S^m$ and has an asymptotic expansion given by (8.3). To this end, we first make note that, by (6.8), (8.12) and (8.16),

$$\tau - \sum_{|\mu|<N_1} \frac{(-i)^{|\mu|}}{\mu!} \partial_x^\mu \partial_\eta^\mu \bar{\sigma} = \sum_{k=0}^\infty T_{N_1}^{(k)} . \qquad (8.20)$$

Let α and β be any two multi-indices. Then, by (8.17) and an integration by parts,

$$\left(D_x^\alpha D_\eta^\beta T_{N_1}^{(k)} \right)(x,\eta)$$

$$= (2\pi)^{-n/2} \int_{\mathbb{R}^n} e^{i\eta \cdot z} z^\beta \left(D_x^\alpha \overline{R_{N_1}^{(k)}} \right)(x,z)dz$$

$$= (1+|\eta|^2)^{-K}(2\pi)^{-n/2} \int_{\mathbb{R}^n} e^{i\eta \cdot z}$$

$$(1-\Delta_z)^K \left\{ z^\beta \left(D_x^\alpha \overline{R_{N_1}^{(k)}} \right)(x,z) \right\} dz , \qquad (8.21)$$

where K is any positive integer. Let $P(D) = (1 - \Delta)^K$. Then, by (8.15), Leibnitz' formulas and an integration by parts,

$$(1 - \Delta_z)^K \left\{ z^\beta \left(D_x^\alpha \overline{R_{N_1}^{(k)}} \right) (x, z) \right\}$$

$$= (1 - \Delta_z)^K N_1 \sum_{|\mu|=N_1} \frac{z^{\mu+\beta}}{\mu!} \int_0^1 (1 - \theta)^{N_1-1} (2\pi)^{-n/2}$$

$$\int_{\mathbb{R}^n} (-i)^{|\alpha|} e^{-iz \cdot \xi} \left(\partial_x^{\alpha+\mu} \bar{\sigma}_k \right) (x + \theta z, \xi) d\xi d\theta$$

$$= N_1 \sum_{|\mu|=N_1} \sum_{|\delta| \leq 2K} \frac{P^{(\delta)}(D) \left(z^{\mu+\beta} \right)}{\mu! \delta!} \int_0^1 (1 - \theta)^{N_1-1} (2\pi)^{-n/2}$$

$$\int_{\mathbb{R}^n} (\dots) d\xi d\theta \,, \tag{8.22}$$

where

$$(\dots)$$

$$= (-i)^{|\alpha|} \sum_{\rho \leq \delta} \binom{\delta}{\rho} \theta^{|\delta-\rho|} \left(D_z^\rho e^{-iz \cdot \xi} \right) \left(D_x^{\delta-\rho} \partial_x^{\alpha+\mu} \bar{\sigma}_k \right) (x + \theta z, \xi) \,. \tag{8.23}$$

Let γ be any multi-index. Then, by (8.22), (8.23), an integration by parts and Leibnitz' formula,

$$(1 - \Delta_z)^K \left\{ z^\beta \left(D_x^\alpha \overline{R_{N_1}^{(k)}} \right) (x, z) \right\}$$

$$= N_1 \sum_{|\mu|=N_1} \sum_{|\delta| \leq 2K} \sum_{\rho \leq \delta} \binom{\delta}{\rho} \frac{P^{(\delta)}(D) \left(z^{\mu+\beta} \right)}{\mu! \delta!} \int_0^1 (1 - \theta)^{N_1-1}$$

$$\theta^{|\delta-\rho|} (2\pi)^{-n/2} (***) d\theta \,, \tag{8.24}$$

where $z^\gamma (***)$ is equal to

$$\int_{\mathbb{R}^n} e^{-iz \cdot \xi} (-i)^{|\alpha|} (-1)^{|\rho|} \sum_{\gamma' \leq \gamma} \binom{\gamma}{\gamma'}$$

$$\left\{ D_\xi^{\gamma'} \left(\xi^\rho D_x^{\delta-\rho} \partial_x^{\alpha+\mu} \bar{\sigma} \right) \right\} (x + \theta z, \xi) \left(D^{\gamma-\gamma'} \varphi_k \right) (\xi) d\xi \,. \tag{8.25}$$

Using the fact that σ is a symbol in S^m, (8.25) and Theorem 6.1 (vi), we can find a constant $C > 0$, depending on γ', ρ, δ, α, μ but not on k, such that

$$|z^\gamma(***)| \leq \int_{W_k} \sum_{\gamma' \leq \gamma} C(1 + |\xi|)^{m+|\rho|-|\gamma'|} 2^{-k|\gamma-\gamma'|} d\xi , \qquad (8.26)$$

where W_0 and $W_k, k = 1, 2, \ldots$, are given by (6.12) and (6.13) respectively. Let M be any positive integer. Then, by (8.26) and Exercise 6.2, there is another constant $C > 0$, depending on M, ρ, δ, α, μ but not on k, such that

$$|(***)| \leq C|z|^{-2M} 2^{k(m+|\rho|-2M+n)} . \qquad (8.27)$$

Hence, by (8.24), (8.27) and Exercise 8.5, there is a constant $C > 0$, depending on α, β, K, M, N_1 but not on k, such that

$$\left| (1 - \Delta_z)^K \left\{ z^\beta \left(D_x^\alpha \overline{R_{N_1}^{(k)}} \right) (x, z) \right\} \right|$$
$$\leq C|z|^{-2M} \left\{ \sum_{|\delta| \leq 2K} |z|^{|\beta|+N_1-|\delta|} \right\} 2^{k(m+2K-2M+n)} . \qquad (8.28)$$

We choose K so large that

$$(1 + |\eta|)^{-2K} \leq (1 + |\eta|)^{m-N-|\beta|} \qquad (8.29)$$

for all $\eta \in \mathbb{R}^n$. Then we choose M so large, say, equal to M', that

$$m + 2K - 2M' + n < 0 . \qquad (8.30)$$

Then we choose N_1 so large that

$$\int_{|z| \leq 1} |z|^{-2M'} \left\{ \sum_{|\delta| \leq 2K} |z|^{|\beta|+N_1-|\delta|} \right\} dz < \infty . \qquad (8.31)$$

Hence, by (8.28) and (8.31), there exists a constant $C_1 > 0$, depending on α, β, K, M', N_1 but not on k, such that

$$\int_{|z|\leq 1} \left| (1-\Delta_z)^K \left\{ z^\beta \left(D_x^\alpha \overline{R_{N_1}^{(k)}}(x,z) \right) \right\} \right| dz \leq C_1 2^{k(m+2K-2M'+n)} .$$

$$(8.32)$$

Now, we choose M so large, say, equal to M'', that

$$m + 2K - 2M'' + n < 0 \qquad (8.33)$$

and

$$\int_{|z|\geq 1} |z|^{-2M''} \left\{ \sum_{|\delta|\leq 2K} |z|^{|\beta|+N_1-|\delta|} \right\} dz < \infty . \qquad (8.34)$$

Hence, by (8.28) and (8.34), there exists a constant $C_2 > 0$, depending on α, β, K, M'', N_1 but not on k, such that

$$\int_{|z|\geq 1} \left| \left(1-\Delta_z\right)^K \left\{ z^\beta \left(D_x^\alpha \overline{R_{N_1}^{(k)}} \right)(x,z) \right\} \right| dz$$
$$\leq C_2\, 2^{k(m+2K-2M''+n)} . \qquad (8.35)$$

Hence, by (8.21), (8.32) and (8.35), we get another constant $C > 0$, depending on α, β, K, M', M'', N_1 but not on k, such that

$$\left| \left(D_x^\alpha D_\eta^\beta T_{N_1}^{(k)} \right)(x,\eta) \right|$$
$$\leq C(1+|\eta|^2)^{-K} \left\{ 2^{k(m+2K-2M'+n)} + 2^{k(m+2K-2M''+n)} \right\} . \qquad (8.36)$$

Hence, for any two multi-indices α and β, we have, by (8.20), (8.29), (8.30), (8.33) and (8.36), a constant $C_{\alpha,\beta} > 0$ for which (8.19) is valid for all x, $\eta \in \mathbb{R}^n$. Therefore the function τ defined by (8.12) is a symbol in S^m and has an asymptotic expansion given by (8.3). That (8.13) is true should be by now obvious. At any rate, it is a simple consequence of Theorem 6.1 (ii), (6.8), (8.4), (8.10) and (8.12).

Exercises

8.1.   Prove that a pseudo-differential operator has a unique formal adjoint.

8.2.   Let σ and τ be any two symbols. Prove that $(T_\sigma^*)^* = T_\sigma$ and $(T_\sigma T_\tau)^* = T_\tau^* T_\sigma^*$.

8.3.   Let $P(x, D) = \sum\limits_{|\alpha| \leq m} a_\alpha(x) D^\alpha$, where the a_α's are in $C^\infty(\mathbb{R}^n)$ and all their partial derivatives are bounded functions on $\mathbb{R}^n$. Compute the symbol of the formal adjoint of $P(x, D)$ directly. Compare the answer with the symbol obtained by Theorem 8.1.

8.4.   Let σ and τ be as in Exercise 7.3. Use Theorem 8.1 to compute the symbols of the formal adjoints of $T_\sigma + T_\tau$, $T_\sigma T_\tau$ and $T_\tau T_\sigma$.

8.5.   In deriving (8.28) from (8.24) and (8.27), we use the fact that there exists a positive constant C, depending on μ, β and K only, such that

$$\sum_{|\delta| \leq 2K} \left| P^{(\delta)}(D)\left(z^{\mu + \beta}\right) \right| \leq C \sum_{|\delta| \leq 2K} |z|^{|\mu| + |\beta| - |\delta|}$$

for all $z \in \mathbb{R}^n$. Prove the fact.

9. THE PARAMETRIX OF AN ELLIPTIC PSEUDO-DIFFERENTIAL OPERATOR

Among all pseudo-differential operators there exists a class of operators which come up frequently in applications and are particularly easy to work with. They are called elliptic operators. They are nice because they have approximate inverses (or parametrices) which are also pseudo-differential operators. Our first task is to make these concepts precise.

A symbol σ in S^m is said to be *elliptic* if there exist positive constants C and R such that

$$|\sigma(x,\xi)| \geq C(1 + |\xi|)^m , \quad |\xi| \geq R .$$

Of course, a pseudo-differential operator T_σ is said to be *elliptic* if its symbol is elliptic.

Theorem 9.1. *Let σ be an elliptic symbol in S^m. Then there exists a symbol τ in S^{-m} such that*

$$T_\tau T_\sigma = I + R \tag{9.1}$$

and

$$T_\sigma T_\tau = I + S \; , \tag{9.2}$$

where R and S are pseudo-differential operators with symbols in $\underset{k \in \mathbb{R}}{\cap} S^k$, and I is the identity operator.

Remark 9.2. In other words, Theorem 9.1 says that if T_σ is an elliptic pseudo-differential operator, then it can be inverted modulo some error terms R and S with symbols in $\underset{k \in \mathbb{R}}{\cap} S^k$. In the theory of regularity of solutions of partial differential equations, operators with symbols in $\underset{k \in \mathbb{R}}{\cap} S^k$ are called *infinitely smoothing* and can be neglected. (See Chapter 14 for a discussion of infinitely smoothing operators and regularity theory.) For these reasons, we call T_τ an approximate inverse, or more often in the literature, a *parametrix* of T_σ.

We first give a proof of (9.1). The idea is to find a sequence of symbols $\tau_j \in S^{-m-j}, j = 0, 1, 2, \ldots$. Let us assume that this has been done. Then, by Theorem 5.10, there exists a symbol $\tau \in S^{-m}$ such that $\tau \sim \sum\limits_{j=0}^{\infty} \tau_j$. Then, by the product formula given in Theorem 7.1, the symbol λ of the product $T_\tau T_\sigma$ is in S^0 and satisfies the following relation:

$$\lambda - \sum_{|\gamma|<N} \frac{(-i)^{|\gamma|}}{\gamma!} (\partial_x^\gamma \sigma)\left(\partial_\xi^\gamma \tau\right) \in S^{-N} \tag{9.3}$$

for every positive integer N. But $\tau \sim \sum\limits_{j=0}^{\infty} \tau_j$ implies that

$$\tau - \sum_{j=0}^{N-1} \tau_j \in S^{-m-N} \tag{9.4}$$

for every positive integer N. Hence, by (9.3) and (9.4),

$$\lambda - \sum_{|\gamma|<N} \frac{(-i)^{|\gamma|}}{\gamma!} (\partial_x^\gamma \sigma) \sum_{j=0}^{N-1} \left(\partial_\xi^\gamma \tau_j\right) \in S^{-N} \tag{9.5}$$

for every positive integer N. But we can write

$$
\sum_{|\gamma|<N} \frac{(-i)^{|\gamma|}}{\gamma!} \sum_{j=0}^{N-1} \left(\partial_\xi^\gamma \tau_j\right) \left(\partial_x^\gamma \sigma\right)
$$

$$
= \tau_0 \sigma + \sum_{l=1}^{N-1} \left\{ \tau_l \sigma + \sum_{\substack{|\gamma|+j=l \\ j<l}} \frac{(-i)^{|\gamma|}}{\gamma!} \left(\partial_\xi^\gamma \tau_j\right) \left(\partial_x^\gamma \sigma\right) \right\}
$$

$$
+ \sum_{\substack{|\gamma|+j\geq N \\ |\gamma|<N,\, j<N}} \frac{(-i)^{|\gamma|}}{\gamma!} \left(\partial_\xi^\gamma \tau_j\right) \left(\partial_x^\gamma \sigma\right) . \tag{9.6}
$$

To simplify (9.6), we choose $\tau_j, j = 0, 1, 2, \ldots$, in the following way:
Define τ_0 by

$$
\tau_0(x,\xi) = \begin{cases} \dfrac{\psi(\xi)}{\sigma(x,\xi)} , & |\xi| > R , \\ 0 , & |\xi| \leq R , \end{cases} \tag{9.7}
$$

where ψ is any function in $C^\infty(\mathbb{R}^n)$ such that $\psi(\xi) = 1$ for $|\xi| \geq 2R$
and $\psi(\xi) = 0$ for $|\xi| \leq R$, and we define τ_l for $l \geq 1$ inductively by

$$
\tau_l = - \left\{ \sum_{\substack{|\gamma|+j=l \\ j<l}} \frac{(-i)^{|\gamma|}}{\gamma!} \left(\partial_x^\gamma \sigma\right) \left(\partial_\xi^\gamma \tau_j\right) \right\} \tau_0 . \tag{9.8}
$$

Then it can be checked easily that $\tau_j \in S^{-m-j}, j = 0, 1, 2, \ldots$. (See
Exercise 9.7.) Now, by (9.7), $\tau_0 \sigma = 1$ for $|\xi| \geq 2R$. The second term
on the right hand side of (9.6) vanishes for $|\xi| \geq 2R$ by (9.7) and
(9.8). As for the third term there, we see easily that $(\partial_\xi^\gamma \tau_j)(\partial_x^\gamma \sigma) \in S^{-N}$ whenever $|\gamma| + j \geq N$. Hence, by (9.6),

$$
\sum_{|\gamma|<N} \frac{(-i)^{|\gamma|}}{\gamma!} \sum_{j=0}^{N-1} \left(\partial_\xi^\gamma \tau_j\right) \left(\partial_x^\gamma \sigma\right) - 1 \in S^{-N} \tag{9.9}
$$

for every positive integer N. Thus, by (9.5) and (9.9), $\lambda - 1 \in S^{-N}$ for every positive integer N. Hence, if we pick R to be the pseudo-differential operator with symbol $\lambda - 1$, then the proof of (9.1) is complete.

By a similar argument, we can find another symbol κ in S^{-m} such that

$$T_\sigma T_\kappa = I + R' , \qquad (9.10)$$

where R' is a pseudo-differential operator with symbol in $\underset{k \in \mathbb{R}}{\cap} S^k$. (See Exercise 9.8.) By (9.1) and (9.10),

$$T_\kappa + R T_\kappa = T_\tau + T_\tau R' .$$

Since $R T_\kappa$ and $T_\tau R'$ are pseudo-differential operators with symbols in $\underset{k \in \mathbb{R}}{\cap} S^k$, it follows that

$$T_\kappa = T_\tau + R'' , \qquad (9.11)$$

where $R'' = T_\tau R' - R T_\kappa$ is another pseudo-differential operator with symbol in $\underset{k \in \mathbb{R}}{\cap} S^k$. Hence, by (9.10) and (9.11),

$$T_\sigma T_\tau = I + S ,$$

where $S = R' - T_\sigma R''$. Since S is a pseudo-differential operator with symbol in $\underset{k \in \mathbb{R}}{\cap} S^k$, it follows that (9.2) is proved.

The following theorem tells us that only elliptic pseudo-differential operators have parametrices.

Theorem 9.3. *Let $\sigma \in S^m$ be such that there exists a $\tau \in S^{-m}$ for which (9.1) or (9.2) is true, where R and S are infinitely smoothing pseudo-differential operators, and I is the identity operator. Then σ is elliptic.*

Remark 9.4. If (9.1) (or (9.2)) is true, then we call T_τ a left (or right) parametrix of T_σ. Thus, a consequence of Theorems 9.1 and

9.3 is that if a pseudo-differential operator T_σ has a left (or right) parametrix T_τ, then T_τ is also a right (or left) parametrix of T_σ.

Proof of Theorem 9.3. Let us first assume that (9.1) is valid and let r be the symbol of R. Then, by Theorem 7.1,

$$1 + r - \sigma\tau = \delta , \qquad (9.12)$$

where δ is some symbol in S^{-1}. Since $\tau \in S^{-m}$, we can find a positive constant C_1 such that

$$|\tau(x,\xi)| \leq C_1(1 + |\xi|)^{-m} , \qquad x, \xi \in \mathbb{R}^n . \qquad (9.13)$$

Thus, by (9.12) and (9.13),

$$|1 + r(x,\xi) - \delta(x,\xi)| \leq C_1|\sigma(x,\xi)|(1 + |\xi|)^{-m} , \qquad x, \xi \in \mathbb{R}^n ,$$

and hence

$$|\sigma(x,\xi)| \geq \frac{1}{C_1}(1 + |\xi|)^m (1 - |\delta(x,\xi)| - |r(x,\xi)|) , \qquad x, \xi \in \mathbb{R}^n . \qquad (9.14)$$

Since $\delta \in S^{-1}$, it follows that there is a positive constant C_2 such that

$$|\delta(x,\xi)| \leq C_2(1 + |\xi|)^{-1} , \qquad x, \xi \in \mathbb{R}^n . \qquad (9.15)$$

Since $r \in \bigcap_{k \in \mathbb{R}} S^k$, we can find a positive constant C_3 such that

$$|r(x,\xi)| \leq C_3(1 + |\xi|)^{-1} , \qquad x, \xi \in \mathbb{R}^n . \qquad (9.16)$$

So, by (9.14)–(9.16),

$$|\sigma(x,\xi)| \geq \frac{1}{C_1}(1 + |\xi|)^m \{1 - (C_2 + C_3)(1 + |\xi|)^{-1}\} , \qquad x, \xi \in \mathbb{R}^n . \qquad (9.17)$$

Now, let R be any positive number such that

$$1 - (C_2 + C_3)(1 + |\xi|)^{-1} < \frac{1}{2} , \qquad |\xi| > R . \qquad (9.18)$$

Then, by (9.17) and (9.18), we get

$$|\sigma(x, \xi)| \geq \frac{1}{2C_1}(1 + |\xi|)^m , \qquad |\xi| > R ,$$

and this completes the proof under the assumption that (9.1) is valid. The proof for the case when (9.2) is valid is similar and hence left as an exercise. See Exercise 9.9.

Exercises

9.1. (i) Let $P(x, D) = \sum\limits_{|\alpha| \leq m} a_\alpha(x) D^\alpha$, where the a_α's are in $C^\infty(\mathbb{R}^n)$ and all their partial derivatives are bounded functions on $\mathbb{R}^n$. We call $\sum\limits_{|\alpha| = m} a_\alpha(x)\xi^\alpha$ the *principal symbol* of $P(x, D)$ and denote it by $P_m(x, \xi)$. Prove that $P(x, D)$ is elliptic if and only if there exist positive constants C and R such that

$$|P_m(x, \xi)| \geq C(1 + |\xi|)^m , \qquad |\xi| \geq R .$$

(ii) Let $P(D) = \sum\limits_{|\alpha| \leq m} a_\alpha D^\alpha$ be a linear partial differential operator with constant coefficients. Let $P_m(\xi)$ be the principal symbol of $P(D)$. Prove that $P(D)$ is elliptic if and only if

$$P_m(\xi) = 0 , \quad \xi \in \mathbb{R}^n \Longrightarrow \xi = 0 .$$

9.2. Let T_σ and T_τ be elliptic pseudo-differential operators. Prove that the product $T_\sigma T_\tau$ is also elliptic.

9.3.   Let T_σ be an elliptic pseudo-differential operator. Prove that the formal adjoint T_σ^* of T_σ is also elliptic.

9.4.   Prove that any two parametrices of an elliptic pseudo-differential operator differ by an infinitely smoothing operator.

9.5.   Is a parametrix of an elliptic pseudo-differential operator elliptic? Explain your answer.

9.6.   Prove that, in the proof of Theorem 9.1, (9.3) and (9.4) imply (9.5).

9.7.   Let $\{\tau_j\}_{j=0}^\infty$ be the sequence of functions defined by (9.7) and (9.8). Prove that τ_j is a symbol in S^{-m-j} for $j = 0, 1, 2, \ldots$.

9.8.   Prove that there exists a symbol κ in S^{-m} such that (9.10) is valid.

9.9.   Let $\sigma \in S^m$ be such that there exists a $\tau \in S^{-m}$ for which (9.2) is true, where S is an infinitely smoothing operator and I is the identity operator. Prove that σ is elliptic.

10. L^p-BOUNDEDNESS OF PSEUDO-DIFFERENTIAL OPERATORS

Let σ be a symbol. Then, by Proposition 5.7, the pseudo-differential operator T_σ maps the Schwartz space $\mathcal{S}$ into $\mathcal{S}$. In fact, the following proposition is true.

Proposition 10.1. *T_σ maps $\mathcal{S}$ continuously into $\mathcal{S}$. More precisely, if $\varphi_k \to 0$ in $\mathcal{S}$, then $T_\sigma \varphi_k \to 0$ in $\mathcal{S}$ as $k \to \infty$.*

To prove Proposition 10.1, we need some preliminary results.

Lemma 10.2. *If $\varphi_k \to 0$ in $\mathcal{S}$ as $k \to \infty$, then $\varphi_k \to 0$ in $L^p(\mathbb{R}^n)$ as $k \to \infty$, for $1 \leq p \leq \infty$.*

Proof: If $\varphi_k \to 0$ in $\mathcal{S}$ as $k \to \infty$, then $\varphi_k \to 0$ uniformly on $\mathbb{R}^n$ as $k \to \infty$. This proves the lemma for $p = \infty$. So, consider $1 \leq p < \infty$. Then, for any positive integer N with $N > \dfrac{n}{p}$, we have

$$(1 + |x|)^N |\varphi_k(x)| \to 0$$

uniformly on $\mathbb{R}^n$ as $k \to \infty$. Hence, for k large enough,

$$|\varphi_k(x)| \leq (1 + |x|)^{-N}, \quad x \in \mathbb{R}^n .$$

Since $(1 + |x|)^{-Np} \in L^1(\mathbb{R}^n)$, it follows from Lebesgue's dominated convergence theorem that

$$\int_{\mathbb{R}^n} |\varphi_k(x)|^p \, dx \to 0$$

as $k \to \infty$. This completes the proof.

Lemma 10.3.   *The Fourier transformation $\mathcal{F}$ maps $\mathcal{S}$ continuously into $\mathcal{S}$. More precisely, if $\varphi_k \to 0$ in $\mathcal{S}$ as $k \to \infty$, then $\hat{\varphi}_k \to 0$ in $\mathcal{S}$ as $k \to \infty$.*

Proof: Let α and β be any two multi-indices. Then, by Proposition 3.2,

$$\left| \xi^\alpha \left(D^\beta \hat{\varphi}_k \right)(\xi) \right| = \left| \left\{ D^\alpha \left((-x)^\beta \varphi_k \right) \right\}^{\hat{}}(\xi) \right|$$

$$\leq (2\pi)^{-n/2} \left\| D^\alpha \left((-x)^\beta \varphi_k \right) \right\|_1 , \quad \xi \in \mathbb{R}^n . \tag{10.1}$$

Since $\varphi_k \to 0$ in $\mathcal{S}$, it follows that $D^\alpha((-x)^\beta \varphi_k) \to 0$ in $\mathcal{S}$ as $k \to \infty$. By Lemma 10.2, $\| D^\alpha((-x)^\beta \varphi_k) \|_1 \to 0$ as $k \to \infty$. By (10.1),

$$\sup_{\xi \in \mathbb{R}^n} \left| \xi^\alpha \left(D^\beta \hat{\varphi}_k \right)(\xi) \right| \to 0$$

as $k \to \infty$. This proves that $\varphi_k \to 0$ in $\mathcal{S}$ as $k \to \infty$.

Proof of Proposition 10.1.   Suppose $\sigma \in S^m$. Then, for any two multi-indices α and β, we have, by (5.12), positive constants $C_{\alpha, \beta, \gamma, \delta}$, depending on α, β, γ and δ only, such that

$$\sup_{x \in \mathbb{R}^n} \left| x^\alpha \left(D^\beta (T_\sigma \varphi_k) \right)(x) \right|$$

$$\leq (2\pi)^{-n/2} \sum_{\gamma \leq \beta} \sum_{\delta \leq \alpha} \binom{\beta}{\gamma} \binom{\alpha}{\delta} C_{\alpha, \beta, \gamma, \delta} \int_{\mathbb{R}^n} (1 + |\xi|)^{m - |\alpha| + |\delta|}$$

$$\left| D_\xi^\delta \left(\xi^\gamma \hat{\varphi}_k(\xi) \right) \right| d\xi . \tag{10.2}$$

Since $\varphi_k \to 0$ in S as $k \to \infty$, it follows easily from Lemma 10.3 that

$$(1 + |\xi|^2)^{(m-|\alpha|+|\delta|)/2} D_\xi^\delta (\xi^\gamma \hat{\varphi}_k(\xi)) \to 0$$

in S as $k \to \infty$. Hence, by Lemma 10.2, the integral on the right hand side of (10.2) goes to zero as $k \to \infty$. This proves that $T_\sigma \varphi_k \to 0$ in S as $k \to \infty$.

The pseudo-differential operator T_σ, initially defined on the Schwartz space S, can be extended to a linear mapping defined on the space S' of tempered distributions. To wit, take a distribution u in S' and define $T_\sigma u$ by

$$(T_\sigma u)(\varphi) = u\overline{(T_\sigma^* \bar{\varphi})}, \quad \varphi \in S, \tag{10.3}$$

where T_σ^* is the formal adjoint of T_σ introduced in Chapter 8.

Proposition 10.4. *T_σ is a linear mapping from S' into S'.*

Proof: Let $u \in S'$. Then, for any sequence $\{\varphi_k\}$ of functions in S converging to zero in S, we have, by (10.3),

$$(T_\sigma u)(\varphi_k) = u\overline{(T_\sigma^* \bar{\varphi}_k)}, \quad k = 1, 2, \ldots. \tag{10.4}$$

By Proposition 10.1, $T_\sigma^* \bar{\varphi}_k \to 0$ in S as $k \to \infty$. Hence, using (10.4) and the fact that u is a tempered distribution, we conclude that $(T_\sigma u)(\varphi_k) \to 0$ as $k \to \infty$. Hence $T_\sigma u \in S'$.

To enquire whether T_σ maps S' continuously into S' or not, we need a notion of convergence in S'.

Definition 10.5. *A sequence of distributions $\{u_k\}$ in S' is said to converge to zero in S'* (denoted by $u_k \to 0$ in S') *if $u_k(\varphi) \to 0$ as $k \to \infty$ for all $\varphi \in S$.*

Proposition 10.6. *T_σ maps S' continuously into S'. More precisely, if $u_k \to 0$ in S' as $k \to \infty$, then $T_\sigma u_k \to 0$ in S' as $k \to \infty$.*

Proof: We need only check Definition 10.5 for the sequence $\{T_\sigma u_k\}$. To do this, let $\varphi \in \mathcal{S}$. Then, using (10.3) and the fact that $u_k \to 0$ in $\mathcal{S}'$ as $k \to \infty$,

$$(T_\sigma u_k)(\varphi) = u_k \overline{(T_\sigma^* \bar{\varphi})} \to 0$$

as $k \to \infty$. Hence $T_\sigma u_k \to 0$ in $\mathcal{S}'$ as $k \to \infty$, and the proof is complete.

Let us recall that, by Proposition 4.5 and Remark 4.6, every function f in $L^p(\mathbb{R}^n)$ is a tempered distribution. Hence, by Proposition 10.4, $T_\sigma f$ is also a tempered distribution. What sort of a tempered distribution is it? We answer this question in the following theorem and Theorem 11.9 in Chapter 11.

Theorem 10.7. *Let σ be a symbol in S^0. Then, for $1 < p < \infty$, $T_\sigma : L^p(\mathbb{R}^n) \to L^p(\mathbb{R}^n)$ is a bounded linear operator.*

The following result plays an important role in our proof of Theorem 10.7. It is a special case of Theorem 2.5 in Hörmander [6]. Its proof is outside the scope of this book and hence omitted.

Theorem 10.8. *Let $m \in C^k(\mathbb{R}^n - \{0\}), k > \frac{n}{2}$, be such that there is a positive constant B for which*

$$|(D^\alpha m)(\xi)| \le B|\xi|^{-|\alpha|}, \quad \xi \ne 0,$$

for all multi-indices α with $|\alpha| \le k$. Then, for $1 < p < \infty$, there is a positive constant C, depending on p and n only, such that

$$\|T\varphi\|_p \le CB\|\varphi\|_p, \quad \varphi \in \mathcal{S},$$

where
$$(T\varphi)(x) = (2\pi)^{-n/2} \int_{\mathbb{R}^n} e^{ix \cdot \xi} m(\xi)\hat{\varphi}(\xi) d\xi, \quad x \in \mathbb{R}^n.$$

Now, we can give a proof of Theorem 10.7.

Proof of Theorem 10.7. Let $\mathbb{Z}^n$ be the set of all n-tuples in $\mathbb{R}^n$ with integer coordinates. We write $\mathbb{R}^n$ as a union of cubes with disjoint interiors, i.e., $\mathbb{R}^n = \underset{m \in \mathbb{Z}^n}{\cup} Q_m$, where Q_m is the cube with center at m, edges of length one and parallel to the coordinate axes. Let Q_0 be the cube with center at the origin. Let η be any function in $C_0^\infty(\mathbb{R}^n)$ such that $\eta(x) = 1$ for all $x \in Q_0$. For $m \in \mathbb{Z}^n$, define σ_m by

$$\sigma_m(x, \xi) = \eta(x - m)\sigma(x, \xi), \quad x, \xi \in \mathbb{R}^n.$$

Obviously, $T_{\sigma_m} = \eta(x - m)T_\sigma$, and

$$\int_{Q_m} |(T_\sigma \varphi)(x)|^p \, dx \le \int_{\mathbb{R}^n} |(T_{\sigma_m} \varphi)(x)|^p \, dx, \quad \varphi \in \mathcal{S}. \tag{10.5}$$

Since $\sigma_m(x, \xi)$ has compact support in x, it follows from Theorem 3.7 and Fubini's theorem that

$$(T_{\sigma_m} \varphi)(x) = (2\pi)^{-n/2} \int_{\mathbb{R}^n} e^{ix \cdot \xi} \sigma_m(x, \xi) \hat{\varphi}(\xi) d\xi$$

$$= (2\pi)^{-n} \int_{\mathbb{R}^n} e^{ix \cdot \xi} \left\{ \int_{\mathbb{R}^n} e^{ix \cdot \lambda} \hat{\sigma}_m(\lambda, \xi) d\lambda \right\} \hat{\varphi}(\xi) d\xi$$

$$= (2\pi)^{-n} \int_{\mathbb{R}^n} e^{ix \cdot \lambda} \left\{ \int_{\mathbb{R}^n} e^{ix \cdot \xi} \hat{\sigma}_m(\lambda, \xi) \hat{\varphi}(\xi) d\xi \right\} d\lambda, \tag{10.6}$$

where

$$\hat{\sigma}_m(\lambda, \xi) = (2\pi)^{-n/2} \int_{\mathbb{R}^n} e^{-i\lambda \cdot x} \sigma_m(x, \xi) dx, \quad \lambda, \xi \in \mathbb{R}^n.$$

Lemma 10.9. *For all multi-indices α and positive integers N, there is a positive constant $C_{\alpha, N}$, depending on α and N only, such that*

$$\left| \left(D_\xi^\alpha \hat{\sigma}_m \right)(\lambda, \xi) \right| \le C_{\alpha, N}(1 + |\xi|)^{-|\alpha|}(1 + |\lambda|)^{-N}, \quad \lambda, \xi \in \mathbb{R}^n.$$

The proof of Lemma 10.9, though easy, will be given later. This lemma and Theorem 10.8 imply that the operator $\varphi \to T_\lambda \varphi$, defined on $\mathcal{S}$ by

$$(T_\lambda \varphi)(x) = (2\pi)^{-n/2} \int_{\mathbb{R}^n} e^{ix\cdot\xi} \hat{\sigma}_m(\lambda, \xi) \hat{\varphi}(\xi) d\xi , \qquad (10.7)$$

can be extended to a bounded linear operator on $L^p(\mathbb{R}^n)$. Moreover, for any positive integer N, there is a positive constant C_N such that

$$\|T_\lambda \varphi\|_p \leq C_N (1+|\lambda|)^{-N} \|\varphi\|_p , \qquad \varphi \in \mathcal{S} . \qquad (10.8)$$

Using (10.6)–(10.8) and Minkowski's inequality in integral form,

$$\|T_{\sigma_m} \varphi\|_p = (2\pi)^{-n/2} \left\{ \int_{\mathbb{R}^n} \left| \int_{\mathbb{R}^n} e^{ix\cdot\lambda} (T_\lambda \varphi)(x) d\lambda \right|^p dx \right\}^{\frac{1}{p}}$$

$$\leq (2\pi)^{-n/2} \int_{\mathbb{R}^n} \left\{ \int_{\mathbb{R}^n} |(T_\lambda \varphi)(x)|^p dx \right\}^{\frac{1}{p}} d\lambda$$

$$= (2\pi)^{-n/2} \int_{\mathbb{R}^n} \|T_\lambda \varphi\|_p d\lambda$$

$$\leq C_N (2\pi)^{-n/2} \left\{ \int_{\mathbb{R}^n} (1+|\lambda|)^{-N} d\lambda \right\} \|\varphi\|_p , \qquad \varphi \in \mathcal{S} .$$

By choosing N sufficiently large, we can get another positive constant C_N such that

$$\|T_{\sigma_m} \varphi\|_p \leq C_N \|\varphi\|_p , \qquad \varphi \in \mathcal{S} . \qquad (10.9)$$

Hence, by (10.5) and (10.9),

$$\int_{Q_m} |(T_\sigma \varphi)(x)|^p dx \leq C_N^p \|\varphi\|_p^p , \qquad \varphi \in \mathcal{S} . \qquad (10.10)$$

Now, we represent T_σ as a singular integral operator. Precisely, we have

Lemma 10.10. *Let* $K(x, z) = (2\pi)^{-n/2} \int_{\mathbb{R}^n} e^{iz \cdot \xi} \sigma(x, \xi) d\xi$ *in the distribution sense. Then*

(i) *for each fixed* $x \in \mathbb{R}^n$, $K(x, \cdot)$ *is a function defined on* $\mathbb{R}^n - \{0\}$,

(ii) *for each sufficiently large positive integer* N, *there is a positive constant* C_N *such that*

$$|K(x, z)| \leq C_N |z|^{-N} , \quad z \neq 0 ,$$

(iii) *for each fixed* $x \in \mathbb{R}^n$ *and* $\varphi \in \mathcal{S}$ *vanishing in a neighborhood of* x,

$$(T_\sigma \varphi)(x) = (2\pi)^{-n/2} \int_{\mathbb{R}^n} K(x, x - z) \varphi(z) dz .$$

Let us assume Lemma 10.10 for a moment. Let Q_m^{**} be the double of Q_m, i.e., Q_m^{**} has the same center as Q_m and edges parallel to the coordinate axes and twice the edge length of Q_m. Let Q_m^* be another cube concentric with Q_m and Q_m^{**} such that $Q_m \subset Q_m^* \subset Q_m^{**}$. (See Figure 2.) Furthermore, we assume that there is a positive number δ such that $|x - z| \geq \delta$ for all $x \in Q_m$ and $z \in \mathbb{R}^n - Q_m^*$.

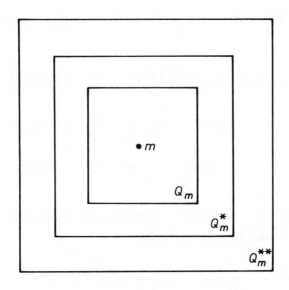

Fig. 2.

Let $\psi \in C_0^\infty(\mathbb{R}^n)$ be such that $0 \le \psi(x) \le 1$ for all $x \in \mathbb{R}^n$, $\text{supp}(\psi) \subseteq Q_m^{**}$, and $\psi(x) = 1$ in a neighborhood of Q_m^*. Write $\varphi = \varphi_1 + \varphi_2$, where $\varphi_1 = \psi\varphi$, and $\varphi_2 = (1 - \psi)\varphi$. Then $T_\sigma\varphi = T_\sigma\varphi_1 + T_\sigma\varphi_2$. Write

$$I_m = \int_{Q_m} |(T_\sigma\varphi)(x)|^p \, dx$$

and

$$J_m = \int_{Q_m} |(T_\sigma\varphi_2)(x)|^p \, dx .$$

Then, for any sufficiently large positive integer N, inequality (10.10) implies that there is a positive constant C_N such that

$$I_m = \int_{Q_m} |(T_\sigma\varphi_1)(x) + (T_\sigma\varphi_2)(x)|^p \, dx$$

$$\le 2^p \int_{Q_m} |(T_\sigma\varphi_1)(x)|^p \, dx + 2^p J_m$$

$$\le 2^p C_N^p \|\varphi_1\|_p^p + 2^p J_m . \tag{10.11}$$

By Lemma 10.10, there is a positive constant C_{2N} such that for all $x \in Q_m$,

$$|(T_\sigma\varphi_2)(x)| = (2\pi)^{-n/2} \left| \int_{\mathbb{R}^n} K(x, x - z)\varphi_2(z) dz \right|$$

$$= (2\pi)^{-n/2} \left| \int_{\mathbb{R}^n - Q_m^*} K(x, x - z)\varphi_2(z) dz \right|$$

$$\le C_{2N} \int_{\mathbb{R}^n - Q_m^*} |x - z|^{-2N} |\varphi_2(z)| \, dz . \tag{10.12}$$

Let $\lambda \ge \sqrt{n} + 1$. Then there exists a positive constant $C_{\lambda,N}$, depending on λ and N only, such that

$$\frac{|x - z|^{-2N}}{(\lambda + |x - z|)^{-2N}} = \frac{(\lambda + |x - z|)^{2N}}{|x - z|^{2N}} \le C_{\lambda,N} \tag{10.13}$$

for all $x \in Q_m$ and $z \in \mathbb{R}^n - Q_m^*$. Hence, by (10.12) and (10.13),

$$|(T_\sigma \varphi_2)(x)|$$
$$\leq C_{2N} C_{\lambda,N} \int_{\mathbb{R}^n - Q_m^*} (\lambda + |x - z|)^{-2N} |\varphi_2(z)| \, dz, \quad x \in Q_m .$$

$$(10.14)$$

Next, we note that, for all $x \in Q_m$ and $z \in \mathbb{R}^n - Q_m^*$,

$$\begin{aligned}
\lambda + |x - z| &= \lambda + |x - m + m - z| \\
&\geq \lambda + |m - z| - |x - m| \\
&\geq \left(\lambda - \frac{\sqrt{n}}{2} \right) + |m - z| \\
&\geq \mu + |m - z| ,
\end{aligned}$$

$$(10.15)$$

where $\mu = \frac{\sqrt{n}}{2} + 1$. By (10.14) and (10.15),

$$|(T_\sigma \varphi_2)(x)|$$
$$\leq C_{2N} C_{\lambda,N} \int_{\mathbb{R}^n - Q_m^*} \frac{(\mu + |x - z|)^{-N} |\varphi_2(z)|}{(\mu + |m - z|)^N} \, dz, \quad x \in Q_m .$$

By Minkowski's inequality in integral form and Hölder's inequality,

$$\left(\int_{Q_m} \left| (T_\sigma \varphi_2)(x) \right|^p dx \right)^{\frac{1}{p}}$$

$$\leq C_{2N} C_{\lambda,N} \left\{ \int_{Q_m} \left| \int_{\mathbb{R}^n - Q_m^*} \frac{(\mu + |x - z|)^{-N} |\varphi_2(z)|}{(\mu + |m - z|)^N} \, dz \right|^p dx \right\}^{\frac{1}{p}}$$

$$\leq C_{2N} C_{\lambda,N} \int_{\mathbb{R}^n - Q_m^*} \left\{ \int_{Q_m} \frac{(\mu + |x - z|)^{-Np} |\varphi_2(z)|^p}{(\mu + |m - z|)^{Np}} \, dx \right\}^{\frac{1}{p}} dz$$

$$= C_{2N} C_{\lambda,N} \int_{\mathbb{R}^n - Q_m^*} \frac{|\varphi_2(z)|}{(\mu + |m - z|)^N}$$

$$\left\{ \int_{Q_m} (\mu + |x - z|)^{-Np} dx \right\}^{\frac{1}{p}} dz$$

$$\leq C_{2N} C_{\lambda,N} \left\{ \int_{\mathbb{R}^n - Q_m^*} (\mu + |m - z|)^{\frac{-Np'}{2}} dz \right\}^{\frac{1}{p'}}$$

$$\left\{ \int_{\mathbb{R}^n - Q_m^*} \frac{|\varphi_2(z)|^p}{(\mu + |m - z|)^{\frac{Np}{2}}} dz \right\}^{\frac{1}{p}} .$$

Hence, for any sufficiently large positive integer N, there is a positive constant $C_{\lambda,N,p}$, depending on λ, N and p only, such that

$$J_m \leq C_{\lambda,N,p} \int_{\mathbb{R}^n - Q_m^*} \frac{|\varphi_2(z)|^p}{(\mu + |m - z|)^{\frac{Np}{2}}} dz . \tag{10.16}$$

By (10.11) and (10.16),

$$\int_{Q_m} |(T_\sigma \varphi)(x)|^p \, dx$$

$$\leq 2^p C_N^p \int_{Q_m^{**}} |\varphi(x)|^p \, dx + 2^p C_{\lambda,N,p} \int_{\mathbb{R}^n - Q_m^*} \frac{|\varphi_2(z)|^p}{(\mu + |m - z|)^{\frac{Np}{2}}} dz .$$

Summing over all m in $\mathbb{Z}^n$, we get a positive constant C, depending only on n, p, N and λ, such that

$$\int_{\mathbb{R}^n} |(T_\sigma \varphi)(x)|^p \, dx$$

$$\leq 2^p C_N^p \sum_{m \in \mathbb{Z}^n} \int_{Q_m^{**}} |\varphi(x)|^p \, dx$$

$$+ 2^p C_{\lambda,N,p} \sum_{m \in \mathbb{Z}^n} \int_{\mathbb{R}^n - Q_m^*} \frac{|\varphi_2(z)|^p}{(\mu + |m - z|)^{\frac{Np}{2}}} dz$$

$$\leq C \int_{\mathbb{R}^n} |\varphi(x)|^p dx + 2^p C_{\lambda,N,p} \sum_{m \in \mathbb{Z}^n} \int_{\mathbb{R}^n - Q_m} \frac{|\varphi_2(z)|^p}{(\mu + |m - z|)^{\frac{Np}{2}}} dz$$

$$= C \int_{\mathbb{R}^n} |\varphi(x)|^p dx + 2^p C_{\lambda,N,p} \sum_{m \in \mathbb{Z}^n} \sum_{l \neq m} \int_{Q_l} \frac{|\varphi_2(z)|^p}{(\mu + |m - z|)^{\frac{Np}{2}}} dz .$$

$$\tag{10.17}$$

But using the same argument as in the derivation of (10.15), we get

$$\mu + |m - z| \geq 1 + |m - l| \qquad (10.18)$$

for all $z \in Q_l$ and $l \neq m$. By (10.18),

$$\sum_{m \in \mathbb{Z}^n} \sum_{l \neq m} \int_{Q_l} \frac{|\varphi_2(z)|^p}{(\mu + |m - l|)^{\frac{Np}{2}}} dz$$

$$\leq \sum_{m \in \mathbb{Z}^n} \sum_{l \neq m} \frac{1}{(1 + |m - l|)^{\frac{Np}{2}}} \int_{Q_l} |\varphi_2(z)|^p dz$$

$$\leq \sum_{m \in \mathbb{Z}^n} \sum_{l \in \mathbb{Z}^n} \frac{1}{(1 + |m - l|)^{\frac{Np}{2}}} \int_{Q_l} |\varphi_2(z)|^p dz$$

$$= \sum_{l \in \mathbb{Z}^n} \int_{Q_l} |\varphi_2(z)|^p dz \sum_{m \in \mathbb{Z}^n} \frac{1}{(1 + |m - l|)^{\frac{Np}{2}}}$$

$$= \sum_{l \in \mathbb{Z}^n} \int_{Q_l} |\varphi_2(z)|^p dz \sum_{m \in \mathbb{Z}^n} \frac{1}{(1 + |m|)^{\frac{Np}{2}}}$$

$$= \sum_{m \in \mathbb{Z}^n} \frac{1}{(1 + |m|)^{\frac{Np}{2}}} \int_{\mathbb{R}^n} |\varphi_2(z)|^p dz . \qquad (10.19)$$

Hence, by (10.17) and (10.19),

$$\int_{\mathbb{R}^n} |(T_\sigma \varphi)(x)|^p dx$$

$$\leq \left\{ C + 2^p C_{\lambda, N, p} \sum_{m \in \mathbb{Z}^n} \frac{1}{(1 + |m|)^{\frac{Np}{2}}} \right\} \int_{\mathbb{R}^n} |\varphi(x)|^p dx .$$

Since $\mathcal{S}$ is dense in $L^p(\mathbb{R}^n)$ by Remark 2.10, it follows that T_σ can be extended to a bounded linear operator on $L^p(\mathbb{R}^n)$.

Remark 10.11. We leave it as an exercise to prove that the bounded extension coincides with $T_\sigma : \mathcal{S}' \to \mathcal{S}'$ restricted to the space $L^p(\mathbb{R}^n)$. See Exercise 10.1.

We now come to the proofs of Lemmas 10.9 and 10.10.

Proof of Lemma 10.9.   Let β be an arbitrary multi-index. Then, by integration by parts and Leibnitz' formula,

$$(-i\lambda)^\beta \left(D_\xi^\alpha \hat{\sigma}_m\right)(\lambda, \xi)$$

$$= (-i\lambda)^\beta D_\xi^\alpha (2\pi)^{-n/2} \int_{\mathbb{R}^n} e^{-ix\cdot\lambda} \sigma_m(x, \xi) dx$$

$$= (-i\lambda)^\beta D_\xi^\alpha (2\pi)^{-n/2} \int_{\mathbb{R}^n} e^{-ix\cdot\lambda} \eta(x - m)\sigma(x, \xi) dx$$

$$= D_\xi^\alpha (2\pi)^{-n/2} \int_{\mathbb{R}^n} \left\{\partial_x^\beta e^{-ix\cdot\lambda}\right\} \eta(x - m)\sigma(x, \xi) dx$$

$$= (2\pi)^{-n/2} \int_{\mathbb{R}^n} \left\{\partial_x^\beta e^{-ix\cdot\lambda}\right\} \eta(x - m) \left(D_\xi^\alpha \sigma\right)(x, \xi) dx$$

$$= (-1)^{|\beta|}(2\pi)^{-n/2} \int_{\mathbb{R}^n} e^{-ix\cdot\lambda} \partial_x^\beta \left\{\eta(x - m) \left(D_\xi^\alpha \sigma\right)(x, \xi)\right\} dx$$

$$= (-1)^{|\beta|}(2\pi)^{-n/2} \sum_{\gamma \le \beta} \binom{\beta}{\gamma} \int_{\mathbb{R}^n} e^{-ix\cdot\lambda} \left(\partial_x^\gamma \eta\right)(x - m)$$
$$\left(\partial_x^{\beta-\gamma} D_\xi^\alpha \sigma\right)(x, \xi) dx .$$

Using the properties of η and the fact that $\sigma \in S^0$, we can find a positive constant $C_{\alpha, \beta}$, depending on α and β only, such that

$$\left|(-i\lambda)^\beta \left(D_\xi^\alpha \hat{\sigma}_m\right)(\lambda, \xi)\right| \le C_{\alpha, \beta}(1 + |\xi|)^{-|\alpha|} , \quad \lambda, \xi \in \mathbb{R}^n .$$

The lemma follows easily from this estimate.

Proof of Lemma 10.10.   Let α be an arbitrary multi-index with length greater than n. Then

$$(-iz)^\alpha K(x, z) = (2\pi)^{-n/2} \int_{\mathbb{R}^n} e^{i\xi\cdot z} \left(\partial_\xi^\alpha \sigma\right)(x, \xi) d\xi \qquad (10.20)$$

in the distribution sense. Since $\sigma \in S^0$, it follows from (10.20) and Proposition 3.3 that $(iz)^\alpha K(x, z)$ is a continuous function on $\mathbb{R}^n$ and there is a positive constant C_α such that

$$|z^\alpha| |K(x, z)| \le C_\alpha$$

for all $x, z \in \mathbb{R}^n$. Hence part (i) follows immediately and part (ii) follows if we use the inequality in Exercise 6.2. To prove part (iii), we define the tempered distribution L_x by

$$L_x(\psi) = \int_{\mathbb{R}^n} \sigma(x, \xi)\psi(\xi)d\xi , \quad \psi \in \mathcal{S} .$$

Then, by the definition of a pseudo-differential operator, Proposition 3.4 and the definition of the Fourier transform of a tempered distribution,

$$\begin{aligned}
(T_\sigma\varphi)(x) &= (2\pi)^{-n/2} \int_{\mathbb{R}^n} e^{ix\cdot\xi}\sigma(x, \xi)\hat{\varphi}(\xi)d\xi \\
&= (2\pi)^{-n/2} L_x(M_x\hat{\varphi}) \\
&= (2\pi)^{-n/2} L_x\left((T_x\varphi)^{\hat{}}\right) \\
&= (2\pi)^{-n/2} \hat{L}_x(T_x\varphi) .
\end{aligned} \tag{10.21}$$

By part (i),

$$\hat{L}_x(\psi) = \int_{\mathbb{R}^n} K(x, -z)\psi(z)dz \tag{10.22}$$

for all $\psi \in \mathcal{S}$ vanishing in a neighborhood of the origin. Hence, by (10.21) and (10.22),

$$\begin{aligned}
(T_\sigma\varphi)(x) &= (2\pi)^{-n/2} \int_{\mathbb{R}^n} K(x, -z)(T_x\varphi)(z)dz \\
&= (2\pi)^{-n/2} \int_{\mathbb{R}^n} K(x, -z)\varphi(x + z)dz \\
&= (2\pi)^{-n/2} \int_{\mathbb{R}^n} K(x, x - z)\varphi(z)dz
\end{aligned}$$

and the proof is complete.

Exercises

10.1.   Let $\sigma \in S^0$ and $1 < p < \infty$. Then we have shown in the proof of Theorem 10.7 that there is a positive constant C such that

$$\|T_\sigma \varphi\|_p \leq C\|\varphi\|_p , \quad \varphi \in \mathcal{S} .$$

Hence T_σ can be extended to a unique bounded linear operator from $L^p(\mathbb{R}^n)$ into $L^p(\mathbb{R}^n)$. Prove that the extension coincides with $T_\sigma :$ $\mathcal{S}' \to \mathcal{S}'$ restricted to $L^p(\mathbb{R}^n)$.

10.2.   Let $\sigma(x, \xi) \in S^0$ be a nonzero symbol which is independent of $\xi \in \mathbb{R}^n$. Prove that the bounded linear operator $T_\sigma : L^p(\mathbb{R}^n) \to L^p(\mathbb{R}^n), 1 < p < \infty$, is not compact.

10.3.   Show that the limit in operator norm of a sequence of pseudo-differential operators $T_{\sigma_k} : L^2(\mathbb{R}^n) \to L^2(\mathbb{R}^n)$, where $\sigma_k \in S^0$, need not be a pseudo-differential operator.

11.   THE SOBOLEV SPACES

$$H^{s,p}, \quad -\infty < s < \infty, \ 1 \leq p < \infty$$

Theorem 10.7 tells us that $T_\sigma : L^p(\mathbb{R}^n) \to L^p(\mathbb{R}^n)$ is a bounded linear operator for $1 < p < \infty$ if σ is a symbol in S^0. In order to find an analogue of Theorem 10.7 for an arbitrary symbol in S^m, we need to introduce a family of spaces of tempered distributions.

For $-\infty < s < \infty$, we denote by J_s the pseudo-differential operator of which the symbol $\sigma_s(\xi)$ is given by

$$\sigma_s(\xi) = \left(1 + |\xi|^2\right)^{-s/2}, \quad \xi \in \mathbb{R}^n .$$

It should be noted that the symbol of J_s is in S^{-s}. (See Example 5.4.) The operator J_s is often called the *Bessel potential of order s*.

It is an easy exercise to prove that for any $u \in \mathcal{S}'$, the product $\sigma_s u$ of σ_s and u defined by

$$(\sigma_s u)(\varphi) = u(\sigma_s \varphi) , \quad \varphi \in \mathcal{S} ,$$

is also in $\mathcal{S}'$. See Exercise 11.1.

The following proposition is an easy consequence of (10.3) and the definition of J_s, and the proof is left as an exercise. See Exercise 11.2.

Proposition 11.1. $J_s u = \mathcal{F}^{-1} \sigma_s \mathcal{F} u$, $u \in \mathcal{S}'$.

An easy corollary of Proposition 11.1 is the following proposition. Its proof is also left as an exercise. See Exercise 11.3.

Proposition 11.2.   *Let $u \in \mathcal{S}'$. Then*

(i) $J_s J_t u = J_{s+t} u$,

(ii) $J_0 u = u$.

For $-\infty < s < \infty$ and $1 \le p < \infty$, we define $H^{s,p}$ to be the set of all tempered distributions u for which $J_{-s} u$ is a function in $L^p(\mathbb{R}^n)$. It is obvious that $H^{s,p}$ is a vector space. It can be made into a normed vector space if we equip it with the norm $\| \ \|_{s,p}$, where

$$\|u\|_{s,p} = \|J_{-s}u\|_p , \quad u \in H^{s,p} .$$

We usually call $H^{s,p}$ the *L^p-Sobolev space of order s.* It is obvious that $H^{0,p} = L^p(\mathbb{R}^n)$.

Theorem 11.3.   *$H^{s,p}$ is a Banach space with respect to the norm $\| \ \|_{s,p}$.*

Proof: We need only prove completeness. To do this, let $\{u_k\}$ be a Cauchy sequence in $H^{s,p}$. Then, by the definition of $H^{s,p}$, the sequence $\{J_{-s}u_k\}$ is a Cauchy sequence in $L^p(\mathbb{R}^n)$. Since $L^p(\mathbb{R}^n)$ is complete, it follows that there exists a function u in $L^p(\mathbb{R}^n)$ such that

$$J_{-s}u_k \to u \quad \text{in} \quad L^p(\mathbb{R}^n) \tag{11.1}$$

as $k \to \infty$. Let $v = J_s u$. Then, by Proposition 11.2, $J_{-s}v = u$. Hence $J_{-s}v$ is in $L^p(\mathbb{R}^n)$, i.e., $v \in H^{s,p}$. That $u_k \to v$ in $H^{s,p}$ as $k \to \infty$ is an immediate consequence of (11.1).

Proposition 11.4.   *J_t is an isometry of $H^{s,p}$ onto $H^{s+t,p}$. More precisely,*

$$\|J_t u\|_{s+t,p} = \|u\|_{s,p} , \quad u \in H^{s,p} . \tag{11.2}$$

Proof: Let $u \in H^{s,p}$. Then, by Proposition 11.2,

$$\|J_t u\|_{s+t,p} = \|J_{-s-t} J_t u\|_p = \|J_{-s} u\|_p = \|u\|_{s,p} .$$

Let $v \in H^{s+t,p}$. Then, by Proposition 11.2, $J_{-t} v \in H^{s,p}$ and $J_t J_{-t} v = v$. This proves that J_t is onto.

Theorem 11.5. *Let* $1 \le p < \infty$ *and* $s \le t$. *Then* $H^{t,p} \subseteq H^{s,p}$, *and*

$$\|u\|_{s,p} \le \|u\|_{t,p} , \quad u \in H^{t,p} .$$

Theorem 11.5 is usually known as the *Sobolev embedding theorem*. To prove Theorem 11.5, we use a technical result which gives us an explicit formula for the inverse Fourier transform of the function $(1 + |\xi|^2)^{-s/2}$, where $\xi \in \mathbb{R}^n$ and $s > 0$, in the distribution sense.

Proposition 11.6. *Let* $s > 0$. *If we define the function* G_s *on* $\mathbb{R}^n$ *by*

$$G_s(x) = \frac{1}{2^{s/2} \Gamma\left(\frac{s}{2}\right)} \int_0^\infty e^{-\frac{r}{2}} e^{\frac{-|x|^2}{2r}} r^{-(n-s)/2} \frac{dr}{r} , \quad x \in \mathbb{R}^n ,$$

then

   (i) $G_s \in L^1(\mathbb{R}^n)$,
   (ii) $\|G_s\|_1 = (2\pi)^{n/2}$,
   (iii) $\hat{G}_s(\xi) = \left(1 + |\xi|^2\right)^{-s/2}$, $\quad \xi \in \mathbb{R}^n$.

Proof: By Fubini's theorem,

$$\int_{\mathbb{R}^n} |G_s(x)| dx = \int_{\mathbb{R}^n} G_s(x) dx$$

$$= \frac{1}{2^{s/2} \Gamma\left(\frac{s}{2}\right)} \int_{\mathbb{R}^n} \left\{ \int_0^\infty e^{-\frac{r}{2}} e^{-\frac{|x|^2}{2r}} r^{-(n-s)/2} \frac{dr}{r} \right\} dx$$

$$= \frac{1}{2^{s/2} \Gamma\left(\frac{s}{2}\right)} \int_0^\infty e^{-\frac{r}{2}} r^{-(n-s)/2} \left\{ \int_{\mathbb{R}^n} e^{-\frac{|x|^2}{2r}} dx \right\} \frac{dr}{r}$$

$$= \frac{(2\pi)^{n/2}}{2^{s/2} \Gamma\left(\frac{s}{2}\right)} \int_0^\infty e^{-\frac{r}{2}} r^{\frac{s}{2}} \frac{dr}{r} \tag{11.3}$$

if we recall that

$$\int_{\mathbb{R}^n} e^{-\frac{|x|^2}{2r}}\, dx = (2\pi r)^{n/2} \ .$$

But, for any $\varepsilon > 0$ and $a > 0$, we have

$$\varepsilon^{-a}\Gamma(a) = \int_0^\infty e^{-\varepsilon r} r^a \frac{dr}{r} \ . \tag{11.4}$$

Putting $\varepsilon = \frac{1}{2}$ and $a = \frac{s}{2}$ in (11.4), we get

$$\int_0^\infty e^{-\frac{r}{2}} r^{\frac{s}{2}} \frac{dr}{r} = 2^{\frac{s}{2}} \Gamma\left(\frac{s}{2}\right) \ . \tag{11.5}$$

Hence, by (11.3) and (11.5),

$$\int_{\mathbb{R}^n} |G_s(x)|\, dx = (2\pi)^{n/2} \ .$$

This proves parts (i) and (ii). To prove part (iii), let $\varphi \in \mathcal{S}$. Then, by Proposition 3.6,

$$\int_{\mathbb{R}^n} \hat{G}_s(\xi)\varphi(\xi)\, d\xi = \int_{\mathbb{R}^n} G_s(\xi)\hat{\varphi}(\xi)\, d\xi$$

$$= \frac{1}{2^{s/2}\Gamma\left(\frac{s}{2}\right)} \int_{\mathbb{R}^n} \left\{\int_0^\infty e^{-\frac{r}{2}} e^{-\frac{|\xi|^2}{2r}} r^{-(n-s)/2} \frac{dr}{r}\right\} \hat{\varphi}(\xi)\, d\xi \ .$$

Using Fubini's theorem, we get

$$\int_{\mathbb{R}^n} \hat{G}_s(\xi)\varphi(\xi)\, d\xi$$

$$= \frac{1}{2^{s/2}\Gamma\left(\frac{s}{2}\right)} \int_0^\infty e^{-\frac{r}{2}} r^{-(n-s)/2} \left\{\int_{\mathbb{R}^n} \hat{\varphi}(\xi) e^{-\frac{|\xi|^2}{2r}}\, d\xi\right\} \frac{dr}{r} \ . \tag{11.6}$$

But, by Proposition 3.6 again, we get

$$\int_{\mathbb{R}^n} \hat{\varphi}(\xi) e^{-\frac{|\xi|^2}{2r}}\, d\xi = \int_{\mathbb{R}^n} \varphi(\xi)\hat{\psi}(\xi)\, d\xi \ , \tag{11.7}$$

where

$$\psi(x) = e^{-\frac{|x|^2}{2r}} , \quad x \in \mathbb{R}^n .$$

By Propositions 3.4 (iii) and 3.5,

$$\hat{\psi}(\xi) = r^{n/2} e^{-r|\xi|^2/2} , \quad \xi \in \mathbb{R}^n .$$

Hence, by (11.7),

$$\int_{\mathbb{R}^n} \hat{\varphi}(\xi) e^{-\frac{|\xi|^2}{2r}} d\xi = r^{\frac{n}{2}} \int_{\mathbb{R}^n} \varphi(\xi) e^{-\frac{r|\xi|^2}{2}} d\xi . \tag{11.8}$$

Therefore, by (11.6) and (11.8),

$$\int_{\mathbb{R}^n} \hat{G}_s(\xi)\varphi(\xi)d\xi = \frac{1}{2^{s/2}\Gamma(\frac{s}{2})} \int_0^\infty e^{-\frac{r}{2}} r^{\frac{s}{2}} \left\{ \int_{\mathbb{R}^n} \varphi(\xi) e^{-\frac{r|\xi|^2}{2}} d\xi \right\} \frac{dr}{r}.$$

Using Fubini's theorem again, we get

$$\int_{\mathbb{R}^n} \hat{G}_s(\xi)\varphi(\xi)d\xi$$

$$= \frac{1}{2^{s/2}\Gamma\left(\frac{s}{2}\right)} \int_{\mathbb{R}^n} \varphi(\xi) \left\{ \int_0^\infty e^{-\frac{r}{2}} r^{\frac{s}{2}} e^{\frac{-r|\xi|^2}{2}} \frac{dr}{r} \right\} d\xi$$

$$= \frac{1}{2^{s/2}\Gamma\left(\frac{s}{2}\right)} \int_{\mathbb{R}^n} \varphi(\xi) \left\{ \int_0^\infty e^{-\frac{r}{2}(1+|\xi|^2)} r^{\frac{s}{2}} \frac{dr}{r} \right\} d\xi .$$

Putting $\varepsilon = (1+|\xi|^2)/2$ and $a = \frac{s}{2}$ in (11.4), we get

$$\int_{\mathbb{R}^n} \hat{G}_s(\xi)\varphi(\xi)d\xi$$

$$= \frac{1}{2^{s/2}\Gamma\left(\frac{s}{2}\right)} \Gamma\left(\frac{s}{2}\right) 2^{s/2} \int_{\mathbb{R}^n} \varphi(\xi) \left(1+|\xi|^2\right)^{-s/2} d\xi$$

$$= \int_{\mathbb{R}^n} \left(1+|\xi|^2\right)^{-s/2} \varphi(\xi)d\xi . \tag{11.9}$$

Hence, by Lemma 5.6 and (11.9), we can conclude that

$$\hat{G}_s(\xi) = \left(1+|\xi|^2\right)^{-s/2} , \quad \xi \in \mathbb{R}^n.$$

The following consequence of Proposition 11.6 will be useful to us.

Proposition 11.7. *Let $s \geq 0$ and $1 \leq p < \infty$. Then*

$$\|J_s u\|_p \leq \|u\|_p , \quad u \in L^p(\mathbb{R}^n). \tag{11.10}$$

Proof: Let $\varphi \in \mathcal{S}$. Then, by the definition of J_s,

$$(J_s\varphi)\widehat{}\,(\xi) = \left(1 + |\xi|^2\right)^{-s/2} \hat{\varphi}(\xi) , \quad \xi \in \mathbb{R}^n.$$

On the other hand, by Propositions 3.1 and 11.6,

$$
\begin{aligned}
(G_s * \varphi)\widehat{}\,(\xi) &= (2\pi)^{n/2} \hat{G}_s(\xi) \hat{\varphi}(\xi) \\
&= (2\pi)^{n/2} \left(1 + |\xi|^2\right)^{-s/2} \hat{\varphi}(\xi) , \quad \xi \in \mathbb{R}^n .
\end{aligned}
$$

Hence, for all $\varphi \in \mathcal{S}$,

$$J_s\varphi = (2\pi)^{-n/2} \left(G_s * \varphi\right) ,$$

and, by Proposition 2.1,

$$\|J_s\varphi\|_p \leq (2\pi)^{-n/2} \|G_s\|_1 \|\varphi\|_p = \|\varphi\|_p .$$

Since $\mathcal{S}$ is dense in $L^p(\mathbb{R}^n)$ for $1 \leq p < \infty$ by Remark 2.10, it follows that J_s can be extended to a bounded linear operator on $L^p(\mathbb{R}^n)$ satisfying (11.10).

Remark 11.8. As in Remark 10.13, we leave it as an exercise to prove that the bounded extension coincides with $J_s : \mathcal{S}' \to \mathcal{S}'$ restricted to the space $L^p(\mathbb{R}^n)$.

Proof of Theorem 11.5. Let $u \in H^{t,p}$. Then, by the definition of $H^{t,p}$, we have $J_{-t}u \in L^p(\mathbb{R}^n)$. Hence, by Proposition 11.2, $J_{-s}u = J_{t-s}J_{-t}u$. By the definition of $H^{s,p}$ and Proposition 11.7, we get

$$\|u\|_{s,p} = \|J_{-s}u\|_p = \|J_{t-s}J_{-t}u\|_p \leq \|J_{-t}u\|_p = \|u\|_{t,p} ,$$

and hence Theorem 11.5 follows.

We can now give a more precise result generalizing Theorem 10.7.

Theorem 11.9. *Let σ be a symbol in S^m. Then $T_\sigma : H^{s,p} \to H^{s-m,p}$ is a bounded linear operator for $-\infty < s < \infty$ and $1 < p < \infty$.*

We first prove a simplified version of Theorem 11.9.

Theorem 11.10. *For any symbol σ in $S^m, T_\sigma : H^{m,p} \to H^{0,p}$ is a bounded linear operator for $1 < p < \infty$.*

Proof: Consider the bounded linear operators

$$J_{-s} : H^{s,p} \to H^{0,p} \ ,$$
$$T_\sigma J_m : H^{0,p} \to H^{0,p} \ ,$$

and

$$J_{s-m} : H^{0,p} \to H^{s-m,p} \ .$$

The first and the third operators are bounded by (11.2), and the second operator is bounded by Theorem 10.7. Hence the product $J_{s-m} T_\sigma J_{m-s}$ is a bounded linear operator from $H^{s,p}$ into $H^{s-m,p}$.

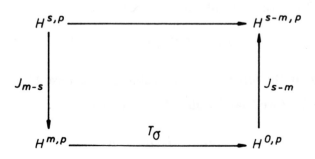

Fig. 3.

Also, by Proposition 11.4, the operators J_{m-s} and J_{s-m} in Figure 3 are isometric and onto. Hence $T_\sigma : H^{m,p} \to H^{0,p}$ must be a bounded linear operator.

Proof of Theorem 11.9.   We begin by making note that $J_{m-s}T_\sigma$ is a pseudo-differential operator with symbol in S^s. Hence, by Theorem 11.10, we can find a positive constant C such that

$$\|T_\sigma u\|_{s-m,\,p} = \|J_{m-s}T_\sigma u\|_p \leq C\|u\|_{s,\,p}$$

for all $u \in H^{s,p}$.

Exercises

11.1.   Let $s \in (-\infty, \infty)$ and σ_s be the function on $\mathbb{R}^n$ defined by

$$\sigma_s(\xi) = (1 + |\xi|^2)^{-s/2} , \quad \xi \in \mathbb{R}^n .$$

Prove that if $u \in \mathcal{S}'$, then the product $\sigma_s u$ of σ_s and u defined by

$$(\sigma_s u)(\varphi) = u(\sigma_s \varphi) , \quad \varphi \in \mathcal{S} ,$$

is also in $\mathcal{S}'$.

11.2.   Prove that for all $s \in (-\infty, \infty)$,

$$J_s u = \mathcal{F}^{-1}\sigma_s \mathcal{F}u , \quad u \in \mathcal{S}' ,$$

where σ_s is the function on $\mathbb{R}^n$ defined in Exercise 11.1.

11.3.   (i) Prove that for all $s, t \in (-\infty, \infty)$,

$$J_s J_t u = J_{s+t}u , \quad u \in \mathcal{S}' .$$

(ii) Prove that $J_0 u = u$ for all $u \in \mathcal{S}'$.

11.4.   (i) Prove that for all $s \geq 0$,

$$H^{s,2} = \left\{ u \in L^2(\mathbb{R}^n) : \int_{\mathbb{R}^n} \left(1 + |\xi|^2\right)^s |\hat{u}(\xi)|^2 d\xi < \infty \right\} .$$

(ii) Prove that for all $s \geq 0$,

$$\|u\|_{s,2} = \left\{ \int_{\mathbb{R}^n} \left(1 + |\xi|^2\right)^s |\hat{u}(\xi)|^2 d\xi \right\}^{\frac{1}{2}}, \quad u \in H^{s,2} .$$

11.5.   (i) Prove that if $u \in H^{s,2}$, $s > \frac{n}{2} + k$, where k is a nonnegative integer, then u is equal to a C^k function on $\mathbb{R}^n$ almost everywhere.
(ii) Prove that if $u \in H^{s,2}$, $s > \frac{n}{2}$, then u can be modified on a set of measure zero to a continuous function v on $\mathbb{R}^n$ such that $\lim_{|x| \to \infty} v(x) = 0$.

11.6.   (**Erhling's Inequality**)   Prove that if $s < t$, then for any positive number ε, there exists a positive constant C, depending on ε, s and t only, such that

$$\|\varphi\|_{s,2} < \varepsilon \|\varphi\|_{t,2} + C \|\varphi\|_{0,2} , \quad \varphi \in \mathcal{S} .$$

11.7.   Let $s \geq 0$ and $1 \leq p < \infty$. Then we have shown in the proof of Proposition 11.7 that

$$\|J_s \varphi\|_p \leq \|\varphi\|_p , \quad \varphi \in \mathcal{S} .$$

Hence J_s can be extended to a unique bounded linear operator from $L^p(\mathbb{R}^n)$ into $L^p(\mathbb{R}^n)$. Prove that the extension coincides with $J_s : \mathcal{S}' \to \mathcal{S}'$ restricted to $L^p(\mathbb{R}^n)$.

11.8.   Let σ be an elliptic symbol in S^m. Let $u \in L^p(\mathbb{R}^n)$ be a solution of the pseudo-differential equation $T_\sigma u = f$, where $f \in L^p(\mathbb{R}^n)$. Prove that $u \in H^{m,p}$.

11.9.   Let $\sigma \in S^m$, $m > 0$, and $f \in L^p(\mathbb{R}^n)$, $1 < p < \infty$. An *approximate solution* of the pseudo-differential equation $T_\sigma u = f$ on $\mathbb{R}^n$ is a function u in $H^{m,p}$ such that $T_\sigma u = f$ modulo $\cap_{s \in \mathbb{R}} H^{s,p}$, i.e., $T_\sigma u - f \in \cap_{s \in \mathbb{R}} H^{s,p}$. Prove that an approximate solution exists if σ is elliptic.

12. CLOSED LINEAR OPERATORS

In this chapter we give a brief account of the theory of closed linear operators on Banach spaces. The choice of topics is dictated by what we need for the theory of minimal and maximal pseudo-differential operators in the next chapter.

Let X and Y be complex Banach spaces with norms denoted by $\| \ \|_X$ and $\| \ \|_Y$ respectively. We are concerned with linear operators A mapping a dense subspace of X, usually denoted by $\mathcal{D}(A)$, into Y. We call $\mathcal{D}(A)$ the *domain* of the operator A.

Definition 12.1. The operator A is said to be *closed* if for any sequence $\{x_k\}$ of vectors in $\mathcal{D}(A)$ such that $x_k \to x$ in X and $Ax_k \to y$ in Y as $k \to \infty$, we have $x \in \mathcal{D}(A)$ and $Ax = y$.

Definition 12.2. The operator A is said to be *closable* if for any sequence $\{x_k\}$ of vectors in $\mathcal{D}(A)$ such that $x_k \to 0$ in X and $Ax_k \to y$ in Y as $k \to \infty$, we have $y = 0$.

Obviously, a closed linear operator is closable.

Definition 12.3. Let A and B be linear operators from X into Y with domains $\mathcal{D}(A)$ and $\mathcal{D}(B)$ respectively. We call B an *extension* of A if $\mathcal{D}(A) \subseteq \mathcal{D}(B)$ and $Bx = Ax$ for all $x \in \mathcal{D}(A)$.

Proposition 12.4.   *Let A be any linear operator from X into Y with domain $\mathcal{D}(A)$. Then A has a closed extension if and only if A is closable.*

Proof: Let B be a closed extension of A. Let $\{x_k\}$ be a sequence of vectors in $\mathcal{D}(A)$ such that $x_k \to 0$ in X and $Ax_k \to y$ in Y as $k \to \infty$. Since B is an extension of A, it follows that $x_k \in \mathcal{D}(B), x_k \to 0$ in X and $Bx_k \to y$ in Y as $k \to \infty$. Since B is closed, we have $y = 0$. Hence A is closable. Conversely, suppose that A is closable. We define an operator A_0 as follows: $\mathcal{D}(A_0)$ is the set of all vectors $x \in X$ such that there is a sequence $\{x_k\}$ of vectors in $\mathcal{D}(A)$ with the property that $x_k \to x$ in X and $Ax_k \to y$ in Y for some $y \in Y$ as $k \to \infty$. For any $x \in \mathcal{D}(A_0)$, we define $A_0 x$ to be equal to y. We have to check that the definition of A_0 does not depend on the particular choice of the sequence $\{x_k\}$. Indeed, if $\{z_k\}$ is another sequence of vectors in $\mathcal{D}(A)$ such that $z_k \to x$ in X and $Az_k \to w$ for some other $w \in Y$ as $k \to \infty$, then $x_k - z_k \to 0$ in X as $k \to \infty, x_k - z_k \in \mathcal{D}(A)$, and $A(x_k - z_k) \to y - w$ in Y as $k \to \infty$. Since A is closable, it follows that $y - w = 0$, i.e., $y = w$. Obviously, A_0 is an extension of A. It is closed. For let $\{x_k\}$ be a sequence of vectors in $\mathcal{D}(A_0)$ such that $x_k \to x$ in X and $A_0 x_k \to y$ in Y as $k \to \infty$. For each k, there is a sequence $\{x_{k_j}\}$ of vectors in $\mathcal{D}(A)$ such that $x_{k_j} \to x_k$ in X and $Ax_{k_j} \to A_0 x_k$ in Y as $j \to \infty$. Hence, for each k, there is a $z_k \in \mathcal{D}(A)$ such that

$$\|x_k - z_k\|_X < \frac{1}{k} \,,$$

and

$$\|A_0 x_k - A z_k\|_Y < \frac{1}{k} \,.$$

Hence, for each k,

$$\|z_k - x\|_X \leq \|x_k - z_k\|_X + \|x_k - x\|_X < \frac{1}{k} + \|x_k - x\|_X \,.$$

Therefore $\|z_k - x\|_X \to 0$ as $k \to \infty$. Similarly, $\|Az_k - y\|_Y \to 0$ as $k \to \infty$. This proves that $x \in \mathcal{D}(A_0)$ and $A_0 x = y$. Hence A_0 is closed.

Let us study the operator A_0 constructed in the proof of Proposition 12.4.

Proposition 12.5. *A_0 is the smallest closed extension of A. This means that if B is any closed extension of A, then B is an extension of A_0.*

Proof: Let $x \in \mathcal{D}(A_0)$ and $A_0 x = y$. Then, by the definition of A_0, we can find a sequence $\{x_k\}$ of vectors in $\mathcal{D}(A)$ such that $x_k \to x$ in X and $Ax_k \to y$ in Y as $k \to \infty$. Since B is an extension of A, it follows that $x_k \in \mathcal{D}(B), x_k \to x$ in X and $Bx_k \to y$ in Y as $k \to \infty$. Since B is closed, we can conclude that $x \in \mathcal{D}(B)$ and $Bx = y$. This proves that B is an extension of A_0.

Remark 12.6. In view of Proposition 12.5, we call A_0 the *minimal operator* of A.

Let X be any complex Banach space with norm $\| \ \|_X$. We denote by X' the dual space of X. Let us recall that X' is the Banach space of all bounded conjugate linear functionals on X. The norm $\| \ \|_{X'}$ in X' is given by

$$\|f\|_{X'} = \sup_{\substack{x \in X \\ x \neq 0}} \frac{|f(x)|}{\|x\|_X} \ , \quad f \in X' \ .$$

Let X and Y be complex Banach spaces. For any linear operator A from X into Y with domain $\mathcal{D}(A)$ *dense* in X, we define an operator $A^t : Y' \to X'$ as follows:

$\mathcal{D}(A^t)$ is the set of all functionals y' in Y' for which there is a functional x' in X' such that

$$y'(Ax) = x'(x) \ , \quad x \in \mathcal{D}(A) \ . \tag{12.1}$$

Lemma 12.7.   *Let $y' \in Y'$. Then there is at most one $x' \in X'$ for which (12.1) holds.*

By Lemma 12.7, we can define $A^t y'$ to be equal to x' for all $y' \in \mathcal{D}(A^t)$. We call A^t the true adjoint or simply the *adjoint* of A.

Proof of Lemma 12.7.   Let $y' \in Y'$. Suppose x' and z' are functionals in X' for which

$$y'(Ax) = x'(x) , \quad x \in \mathcal{D}(A) ,$$

and

$$y'(Ax) = z'(x) , \quad x \in \mathcal{D}(A) .$$

Obviously, $x' = z'$ on $\mathcal{D}(A)$. Since $\mathcal{D}(A)$ is dense in X, a simple limiting argument will show that $x' = z'$ on X.

Proposition 12.8.   *A^t is a closed linear operator from Y' into X'.*

Proof: Linearity, as usual, is easy to check. To prove that A^t is closed, let $\{y_k'\}$ be a sequence of functionals in $\mathcal{D}(A^t)$ such that $y_k' \to y'$ in Y' and $A^t y_k' \to x'$ in X' as $k \to \infty$. Then, by (12.1) and the definition of A^t,

$$y_k'(Ax) = (A^t y_k')(x)$$

for all $x \in \mathcal{D}(A)$ and $k = 1, 2, \dots$. Let $k \to \infty$. Then

$$y'(Ax) = x'(x) , \quad x \in \mathcal{D}(A) .$$

This proves that $y' \in \mathcal{D}(A^t)$ and $A^t y' = x'$. Therefore A^t is a closed operator.

Another observation about the adjoint of a linear operator we shall use is given in the following proposition:

Proposition 12.9.   *Let A be any linear operator from X into Y with domain $\mathcal{D}(A)$ dense in X. Then, for any extension B of A, the operator A^t is an extension of B^t.*

Proof: Let $y' \in \mathcal{D}(B^t)$. Then we can find a functional x' in X' such that

$$y'(Bx) = x'(x) , \quad x \in \mathcal{D}(B) .$$

Since B is an extension of A, we have

$$y'(Ax) = x'(x) , \quad x \in \mathcal{D}(A) .$$

This proves that $y' \in \mathcal{D}(A^t)$ and $A^t y' = x' = B^t y'$.

Exercises

12.1. Let A be a linear operator from $L^p(\mathbb{R}^n)$ into $L^p(\mathbb{R}^n)$, $1 < p < \infty$, with dense domain $\mathcal{D}(A)$. Let p' be the conjugate index of p, i.e., $\dfrac{1}{p} + \dfrac{1}{p'} = 1$. Prove that $\mathcal{D}(A^t)$ consists of all functions u in $L^{p'}(\mathbb{R}^n)$ for which there exists a function f in $L^{p'}(\mathbb{R}^n)$ such that

$$(u, Av) = (f, v) , \quad v \in \mathcal{D}(A) ,$$

where

$$(g, h) = \int_{\mathbb{R}^n} g(x)\overline{h(x)}dx$$

for all $g \in L^{p'}(\mathbb{R}^n)$ and $h \in L^p(\mathbb{R}^n)$.

12.2. Let Z be a complex Banach space. A subset S of Z' is said to be *total* if

$$\{z \in Z : z'(z) = 0 \quad \text{for all} \quad z' \in S\} = \{0\} .$$

Let A be a linear operator from a complex Banach space into another complex Banach space with dense domain $\mathcal{D}(A)$. Prove that A is closable if and only if the domain $\mathcal{D}(A^t)$ of A^t is a total set.

13. MINIMAL AND MAXIMAL PSEUDO-DIFFERENTIAL OPERATORS

Let σ be a symbol in S^m. Then the pseudo-differential operator T_σ was initially defined on the Schwartz space S and later extended to the space S' of all tempered distributions by using the formal adjoint T_σ^*. It was shown in Theorem 11.9 that $T_\sigma : H^{s,p} \to H^{s-m,p}$ was a bounded linear operator from $-\infty < s < \infty$ and $1 < p < \infty$. As a matter of fact, when $m \geq 0$, the operator T_σ can also be considered as a linear operator from $L^p(\mathbb{R}^n)$ into $L^p(\mathbb{R}^n)$, $1 < p < \infty$, with domain S. We denote this operator simply by T_σ. It is not closed in general. Fortunately, it is closable. Hence, by Theorem 12.4, it has a closed extension.

Proposition 13.1. *The operator T_σ is closable.*

Proof: Let $\{\varphi_k\}$ be a sequence of functions in S such that $\varphi_k \to 0$ and $T_\sigma \varphi_k \to f$ in $L^p(\mathbb{R}^n)$ as $k \to \infty$. Then, for any function ψ in S, we have

$$(T_\sigma \varphi_k, \psi) = (\varphi_k, T_\sigma^* \psi) , \quad k = 1, 2, \dots ,$$

where T_σ^* is the formal adjoint of T_σ. Let $k \to \infty$. Then $(f, \psi) = 0$ for all functions $\psi \in S$. Since S is dense in $L^p(\mathbb{R}^n)$, it follows that

$f = 0$. Hence, by Definition 12.2, T_σ is closable.

Remark 13.2. A consequence of Proposition 13.1 is that the minimal operator $T_{\sigma, 0}$ of T_σ exists. (See Remark 12.6.) Let us recall that the domain $\mathcal{D}(T_{\sigma, 0})$ of $T_{\sigma, 0}$ consists of all functions u in $L^p(\mathbb{R}^n)$ for which a sequence $\{\varphi_k\}$ in $\mathcal{S}$ can be found such that $\varphi_k \to u$ in $L^p(\mathbb{R}^n)$ and $T_\sigma \varphi_k \to f$ in $L^p(\mathbb{R}^n)$ for some $f \in L^p(\mathbb{R}^n)$ as $k \to \infty$. Moreover, $T_{\sigma, 0} u = f$.

Definition 13.3. Let u and f be functions in $L^p(\mathbb{R}^n), 1 < p < \infty$. We say that u lies in $\mathcal{D}(T_{\sigma, 1})$ and $T_{\sigma, 1} u = f$ if and only if

$$(u, T_\sigma^* \varphi) = (f, \varphi) , \quad \varphi \in \mathcal{S} , \tag{13.1}$$

where T_σ^* is the formal adjoint of T_σ.

Proposition 13.4. *Let $u \in \mathcal{D}(T_{\sigma, 1})$. Then $T_{\sigma, 1} u = T_\sigma u$ in the distribution sense.*

Proof: By Definition 13.3,

$$(T_{\sigma, 1} u, \varphi) = (u, T_\sigma^* \varphi) , \quad \varphi \in \mathcal{S} .$$

Hence, by considering u and $T_{\sigma, 1} u$ as tempered distributions, we have

$$(T_{\sigma, 1} u)(\bar{\varphi}) = u\overline{(T_\sigma^* \varphi)} , \quad \varphi \in \mathcal{S} . \tag{13.2}$$

On the other hand, by (10.3),

$$(T_\sigma u)(\bar{\varphi}) = u\overline{(T_\sigma^* \varphi)} , \quad \varphi \in \mathcal{S} . \tag{13.3}$$

Hence, by (13.2) and (13.3), $T_{\sigma, 1} u = T_\sigma u$ in the distribution sense.

Proposition 13.5. $T_{\sigma, 1}$ *is a closed linear operator from $L^p(\mathbb{R}^n)$ into $L^p(\mathbb{R}^n)$ with domain $\mathcal{D}(T_{\sigma, 1})$ containing $\mathcal{S}$.*

Proof: That $S \subseteq \mathcal{D}(T_{\sigma,1})$ is obvious from (13.1) and the definition of the formal adjoint T_σ^*. Linearity is again easy to check. To prove that $T_{\sigma,1}$ is closed, let $\{u_k\}$ be a sequence of functions in $\mathcal{D}(T_{\sigma,1})$ such that $u_k \to u$ in $L^p(\mathbb{R}^n)$ and $T_{\sigma,1}u_k \to f$ in $L^p(\mathbb{R}^n)$ for some u and $f \in L^p(\mathbb{R}^n)$ as $k \to \infty$. Then, by (13.1),

$$(u_k, T_\sigma^*\varphi) = (T_{\sigma,1}u_k, \varphi) \tag{13.4}$$

for all $\varphi \in S$ and $k = 1, 2, \ldots$. Let $k \to \infty$ in (13.4). Then

$$(u, T_\sigma^*\varphi) = (f, \varphi), \quad \varphi \in S .$$

Hence, by Definition 13.3, $u \in \mathcal{D}(T_{\sigma,1})$ and $T_{\sigma,1}u = f$. This proves that $T_{\sigma,1}$ is closed by Definition 12.1.

Proposition 13.6. $S \subseteq \mathcal{D}(T_{\sigma,1}^t)$, where $T_{\sigma,1}^t$ is the adjoint of $T_{\sigma,1}$.

Proof: Since $T_{\sigma,1}$ is a closed linear operator from $L^p(\mathbb{R}^n)$ into $L^p(\mathbb{R}^n), 1 < p < \infty$, with domain containing S, it follows from Proposition 12.8 that $T_{\sigma,1}^t$ is a closed linear operator from $L^{p'}(\mathbb{R}^n)$ into $L^{p'}(\mathbb{R}^n)$, where p' is the conjugate index of p. Let $\psi \in S$. Then, for all functions $u \in \mathcal{D}(T_{\sigma,1})$,

$$(\psi, T_{\sigma,1}u) = (T_\sigma^*\psi, u)$$

by Definition 13.3. Hence, by the definition of $T_{\sigma,1}^t$, $\psi \in \mathcal{D}(T_{\sigma,1}^t)$ and $T_{\sigma,1}^t\psi = T_\sigma^*\psi$.

Proposition 13.7. $T_{\sigma,1}$ is an extension of $T_{\sigma,0}$.

Proof: Let $u \in \mathcal{D}(T_{\sigma,0})$ and $T_{\sigma,0}u = f$. Then, by Remark 13.2, there is a sequence $\{\varphi_k\}$ of functions in S for which $\varphi_k \to u$ and $T_\sigma\varphi_k \to f$ in $L^p(\mathbb{R}^n)$ as $k \to \infty$. Hence, by the definition of T_σ^*, we have

$$(\varphi_k, T_\sigma^*\psi) = (T_\sigma\varphi_k, \psi)$$

for all $\psi \in \mathcal{S}$ and $k = 1, 2, \ldots$. Let $k \to \infty$. Then

$$(u, T_\sigma^* \psi) = (f, \psi) , \quad \psi \in \mathcal{S} .$$

So, by Definition 13.3, $u \in \mathcal{D}(T_{\sigma,1})$ and $T_{\sigma,1} u = f$.

Remark 13.8. Using Propositions 12.9 and 13.7, we see that $T_{\sigma,0}^t$ is an extension of $T_{\sigma,1}^t$. Since, by Proposition 13.6, the domain of $T_{\sigma,1}^t$ contains the space $\mathcal{S}$, it follows that the domain of $T_{\sigma,0}^t$ contains $\mathcal{S}$ as well.

Proposition 13.9. *$T_{\sigma,1}$ is the largest closed extension of T_σ having $\mathcal{S}$ contained in the domain of its adjoint. In other words, if B is any closed extension of T_σ such that $\mathcal{S} \subseteq \mathcal{D}(B^t)$, then $T_{\sigma,1}$ is an extension of B.*

We first prove the following lemma:

Lemma 13.10. *$T_\sigma^t \varphi = T_\sigma^* \varphi$ for all $\varphi \in \mathcal{S}$. In other words, the true and formal adjoints coincide on the space $\mathcal{S}$.*

Proof: Let $\varphi \in \mathcal{S}$. Then, by the definition of T_σ^*, we have

$$(T_\sigma^* \varphi, \psi) = (\varphi, T_\sigma \psi) , \quad \psi \in \mathcal{S} .$$

So, by the definition of T_σ^t and the duality of $L^p(\mathbb{R}^n)$, $\varphi \in \mathcal{D}(T_\sigma^t)$ and $T_\sigma^t \varphi = T_\sigma^* \varphi$.

Proof of Proposition 13.9. Let $u \in \mathcal{D}(B)$. Then, for all $\psi \in \mathcal{S}$, we have $\psi \in \mathcal{D}(B^t)$. Hence, by the definition of B^t,

$$(\psi, Bu) = (B^t \psi, u) . \tag{13.5}$$

Since B is an extension of T_σ, it follows from Proposition 12.9 that T_σ^t is an extension of B^t. Hence, by (13.5),

$$(\psi, Bu) = (T_\sigma^t \psi, u) . \tag{13.6}$$

By Lemma 13.10, $T_\sigma^t = T_\sigma^*$ on $\mathcal{S}$. Hence, by (13.6), we have

$$(\psi, Bu) = (T_\sigma^* \psi, u) \ , \quad \psi \in \mathcal{S} \ .$$

Therefore, by Definition 13.3, we have $u \in \mathcal{D}(T_{\sigma,1})$ and $T_{\sigma,1}u = Bu$.

Remark 13.11. Because of Proposition 13.9, we call $T_{\sigma,1}$ the *maximal operator of T_σ*.

By Proposition 13.7, we know that $T_{\sigma,1}$ is an extension of $T_{\sigma,0}$. The aim of this chapter is to prove that $T_{\sigma,0} = T_{\sigma,1}$ if σ is an elliptic symbol in S^m, $m \geq 0$. (See Chapter 9 for the definition of ellipticity.) We need some preparation.

Theorem 13.12. *Let $m \geq 0$ and σ be an elliptic symbol in S^m. Then $\mathcal{D}(T_{\sigma,0}) = H^{m,p}$.*

To prove Theorem 13.12, we use the following estimate, which is the analogue of the Agmon-Douglis-Nirenberg estimate in [1] for pseudo-differential operators.

Proposition 13.13. *Let $m \geq 0$ and σ be an elliptic symbol in S^m. Then there exist positive constants C_1 and C_2 such that*

$$C_1 \|u\|_{m,p} \leq (\|T_\sigma u\|_{0,p} + \|u\|_{0,p}) \leq C_2 \|u\|_{m,p} \ , \quad u \in H^{m,p} \ .$$

Proof: By Theorems 11.5 and 11.9, there is a positive constant C' such that

$$\|T_\sigma u\|_{0,p} + \|u\|_{0,p} \leq C' \|u\|_{m,p} \ , \quad u \in H^{m,p} \ .$$

Next, by (9.1) in Theorem 9.1, we have

$$u = T_\tau T_\sigma u - Ru \ , \quad u \in H^{m,p} \ , \tag{13.7}$$

where $\tau \in S^{-m}$ and R is a pseudo-differential operator with symbol in $\underset{k \in \mathbb{R}}{\cap} S^k$. Hence it follows from Theorem 11.9 and (13.7) that there is a positive constant C such that

$$C\|u\|_{m,p} \leq (\|T_\sigma u\|_{0,p} + \|u\|_{0,p}) , \quad u \in H^{m,p} .$$

This proves Proposition 13.13.

Proposition 13.14. *S is dense in $H^{s,p}$, $-\infty < s < \infty, 1 < p < \infty$.*

Proof: Let $u \in H^{s,p}$. Then, by the definition of $H^{s,p}, J_{-s}u \in L^p(\mathbb{R}^n)$. Since S is dense in $L^p(\mathbb{R}^n)$ by Remark 2.10, it follows that there is a sequence $\{\varphi_k\}$ of functions in S such that $\varphi_k \to J_{-s}u$ in $L^p(\mathbb{R}^n)$ as $k \to \infty$. Let $\psi_k = J_s\varphi_k, k = 1, 2, \ldots$. By Proposition 5.7, $\psi_k \in S, k = 1, 2, \ldots$. Also, by the definition of $H^{s,p}$ again,

$$\|\psi_k - u\|_{s,p} = \|J_{-s}\psi_k - J_{-s}u\|_p = \|\varphi_k - J_{-s}u\|_p \to 0$$

as $k \to \infty$. This proves that S is dense in $H^{s,p}$.

We can now prove Theorem 13.12.

Proof of Theorem 13.12. Let $u \in H^{m,p}$. Then, by Proposition 13.14, we can find a sequence $\{\varphi_k\}$ of functions in S such that $\varphi_k \to u$ in $H^{m,p}$ as $k \to \infty$. By Propositions 13.13 and 13.14, $\{T_\sigma\varphi_k\}$ and $\{\varphi_k\}$ are Cauchy sequences in $L^p(\mathbb{R}^n)$. Hence $\varphi_k \to u$ and $T_\sigma\varphi_k \to f$ in $L^p(\mathbb{R}^n)$ for some u and f in $L^p(\mathbb{R}^n)$ as $k \to \infty$. Hence, by the definition of $T_{\sigma,0}, u \in \mathcal{D}(T_{\sigma,0})$ and $T_{\sigma,0}u = f$. On the other hand, if $u \in \mathcal{D}(T_{\sigma,0})$, then, by the definition of $T_{\sigma,0}$ again, we can find a sequence $\{\varphi_k\}$ of functions in S for which $\varphi_k \to u$ in $L^p(\mathbb{R}^n)$ and $T_\sigma\varphi_k \to f$ in $L^p(\mathbb{R}^n)$ for some $f \in L^p(\mathbb{R}^n)$ as $k \to \infty$. Hence $\{\varphi_k\}$ and $\{T_\sigma\varphi_k\}$ are Cauchy sequences in $L^p(\mathbb{R}^n)$. So, by Propositions 13.13 and 13.14, $\{\varphi_k\}$ is a Cauchy sequence in $H^{m,p}$. Since $H^{m,p}$ is complete by Theorem 11.3, it follows that $\varphi_k \to v$ in $H^{m,p}$ for some

$v \in H^{m,p}$ as $k \to \infty$. Then, by Theorem 11.5, $\varphi_k \to v$ in $L^p(\mathbb{R}^n)$ as $k \to \infty$. Hence $u = v$ and consequently $u \in H^{m,p}$.

Finally we come to the main result of this chapter.

Theorem 13.15. *Let $m \geq 0$ and σ be an elliptic symbol in S^m. Then $T_{\sigma,0} = T_{\sigma,1}$.*

Proof: Since $T_{\sigma,0}$ is the smallest closed extension of T_σ, it follows from Proposition 13.7 and Theorem 13.12 that it is sufficient to prove that $\mathcal{D}(T_{\sigma,1}) \subseteq H^{m,p}$. Let $u \in \mathcal{D}(T_{\sigma,1})$. Then, by (9.1) in Theorem 9.1,

$$u = T_\tau T_\sigma u - Ru , \qquad (13.8)$$

where $\tau \in S^{-m}$ and R is a pseudo-differential operator with symbol in $\underset{k \in \mathbb{R}}{\cap} S^k$. By Proposition 13.4, $T_{\sigma,1}u = T_\sigma u$ in the distribution sense. Thus, by the definition of $T_{\sigma,1}, T_\sigma u \in L^p(\mathbb{R}^n)$. Since $\tau \in S^{-m}$, it follows from Theorem 11.9 that $T_\tau T_\sigma u \in H^{m,p}$. Since $u \in L^p(\mathbb{R}^n)$ and R has symbol in S^{-m}, it follows from Theorem 11.9 again that $Ru \in H^{m,p}$. Hence, by (13.8), $u \in H^{m,p}$.

Exercises

13.1.   Let σ be any symbol in $S^m, m \leq 0$. Prove that the minimal operator $T_{\sigma,0}$ of $T_\sigma : \mathcal{S} \to \mathcal{S}$ in $L^p(\mathbb{R}^n), 1 < p < \infty$, is a bounded linear operator from $L^p(\mathbb{R}^n)$ into $L^p(\mathbb{R}^n)$.

13.2.   Let σ be any symbol. Consider T_σ as a linear operator from $L^p(\mathbb{R}^n)$ into $L^p(\mathbb{R}^n), 1 < p < \infty$, with dense domain $\mathcal{S}$. Prove that $T_\sigma^t = (T_{\sigma,0})^t$, where $(T_{\sigma,0})^t$ is the adjoint of $T_{\sigma,0}$.

13.3. Let σ and T_σ be as in Exercise 13.2. Prove that $T_\sigma^t = (T_\sigma^*)_1$, where $(T_\sigma^*)_1$ is the maximal operator of the pseudo-differential operator T_σ^*.

13.4. For any closed linear operator A from a complex Banach space into itself with domain $\mathcal{D}(A)$, a subspace $\mathcal{D}$ of $\mathcal{D}(A)$ is called a *core* of the operator A if the minimal operator of the restriction of A to $\mathcal{D}$ is equal to A. Prove that if σ is any elliptic symbol, then $C_0^\infty(\mathbb{R}^n)$ is a core of $T_{\sigma,0}$ and $(T_{\sigma,0})^t$.

14. GLOBAL REGULARITY OF ELLIPTIC PARTIAL DIFFERENTIAL EQUATIONS

Let $P(x, D) = \sum\limits_{|\alpha| \leq m} a_\alpha(x) D^\alpha$ be a linear partial differential operator of order m such that

$$\sup_{x \in \mathbb{R}^n} \left| \left(D^\beta a_\alpha \right)(x) \right| < \infty \,, \quad |\alpha| \leq m \,, \tag{14.1}$$

for all multi-indices β. Then we have observed in Example 5.3 that $P(x, D)$ is a pseudo-differential operator with symbol $P(x, \xi)$ in S^m, where

$$P(x, \xi) = \sum_{|\alpha| \leq m} a_\alpha(x) \xi^\alpha \,.$$

The purpose of this chapter is to use the theory of pseudo-differential operators we have developed to prove the following L^p analogue of a result in Hess and Kato [5].

Theorem 14.1. *Let $P(x, D) = \sum\limits_{|\alpha| \leq m} a_\alpha(x) D^\alpha$ be a linear partial differential operator of order m satisfying (14.1). Let $x_0 \in \mathbb{R}^n$. Suppose that there exist positive constants C_1 and C_2 such that*

$$\left| \sum_{|\alpha| = m} a_\alpha(x_0) \xi^\alpha \right| \geq C_1 |\xi|^m \tag{14.2}$$

and

$$\left| \sum_{|\alpha|=m} (a_\alpha(x) - a_\alpha(x_0)) \, \xi^\alpha \right| \leq C_2 |\xi|^m \qquad (14.3)$$

for all x and $\xi \in \mathbb{R}^n$, and $C_2 < C_1$. If $u \in H^{s,p}$, $P(x,D)u = f$ and $f \in H^{s,p}$, then $u \in H^{s+m,p}$.

Remark 14.2. If $P(x,D)$ is a linear partial differential operator of order m satisfying the hypotheses of Theroem 14.1, and if f is any tempered distribution in $H^{s,p}$, then Theorem 14.1 asserts that any solution u in $H^{s,p}$ of the partial differential equation $P(x,D)u = f$ on $\mathbb{R}^n$ lies in a more "selective" or "regular" space $H^{s+m,p}$. If we recall, by Theorem 11.5, that

$$H^{s+m,p} \subset H^{s+m-1,p} \subset \cdots \subset H^{s,p} \,,$$

then the solution u can be thought of being m steps more "selective" or "regular" than the given source data f defined globally on $\mathbb{R}^n$. For this reason, we call Theorem 14.1 a global regularity theorem.

To prove Theorem 14.1, we use the following lemma:

Lemma 14.3. *Let $P(x,D) = \sum\limits_{|\alpha| \leq m} a_\alpha(x) D^\alpha$ be a linear partial differential operator of order m satisfying the hypotheses of Theorem 14.1. Then there exist positive constants C and R such that*

$$|P(x,\xi)| \geq C(1+|\xi|)^m \,, \quad |\xi| \geq R \,.$$

Remark 14.4. Lemma 14.3 tells us that under the hypotheses of Theorem 14.1, $P(x,D)$ is a pseudo-differential operator with symbol $P(x,\xi)$ in S^m and $P(x,\xi)$ satisfies the ellipticity condition defined in Chapter 9.

Proof of Lemma 14.3. We begin by making note that

$$\left| \sum_{|\alpha|=m} a_\alpha(x)\xi^\alpha \right| = \left| \sum_{|\alpha|=m} (a_\alpha(x) - a_\alpha(x_0))\xi^\alpha + \sum_{|\alpha|=m} a_\alpha(x_0)\xi^\alpha \right|$$

$$\geq \left| \sum_{|\alpha|=m} a_\alpha(x_0)\xi^\alpha \right| - \left| \sum_{|\alpha|=m} (a_\alpha(x) - a_\alpha(x_0))\xi^\alpha \right|$$

$$\geq (C_1 - C_2)|\xi|^m, \quad x, \xi \in \mathbb{R}^n, \tag{14.4}$$

if we make use of (14.2) and (14.3). Next, by (14.4), we can find positive constants C', C'' and R such that

$$\left| \sum_{|\alpha|\leq m} a_\alpha(x)\xi^\alpha \right| = \left| \sum_{|\alpha|=m} a_\alpha(x)\xi^\alpha + \sum_{|\alpha|<m} a_\alpha(x)\xi^\alpha \right|$$

$$\geq \left| \sum_{|\alpha|=m} a_\alpha(x)\xi^\alpha \right| - \left| \sum_{|\alpha|<m} a_\alpha(x)\xi^\alpha \right|$$

$$\geq C'(1+|\xi|)^m - C''(1+|\xi|)^{m-1}$$

$$= (1+|\xi|)^m \left(C' - C''(1+|\xi|)^{-1} \right), \quad |\xi| \geq R. \tag{14.5}$$

Obviously, we can find another positive constant $R_1 \geq R$ such that

$$C' - C''(1+|\xi|)^{-1} \geq \frac{C'}{2}, \quad |\xi| \geq R_1. \tag{14.6}$$

Hence, by (14.5) and (14.6),

$$\left| \sum_{|\alpha|\leq m} a_\alpha(x)\xi^\alpha \right| \geq \frac{C'}{2}(1+|\xi|)^m, |\xi| \geq R_1.$$

This proves Lemma 14.3.

Proof of Theorem 14.1. By Lemma 14.3, $P(x, D)$ is an elliptic pseudo-differential operator with symbol $P(x, \xi)$ in S^m. Hence, by Theorem 9.1, we can find a symbol $\tau \in S^{-m}$ and a pseudo-differential operator R with symbol in $\bigcap_{k \in \mathbb{R}} S^k$ such that

$$T_\tau P(x, D) = I + R. \tag{14.7}$$

Hence, by (14.7), $u = T_\tau f - Ru$. Since $\tau \in S^{-m}$ and $f \in H^{s,p}$, it follows that $T_\tau f \in H^{s+m,p}$. Also, $Ru \in H^{s+m,p}$ because R has symbol in S^{-m} and $u \in H^{s,p}$. Hence, by (14.7), $u \in H^{s+m,p}$.

Exercises

14.1.   Let $P(x, D)$ be a linear partial differential operator of order m satisfying all the hypotheses of Theorem 14.1. Prove that, if $f \in H^{s,p}$, then any solution u in $\bigcup_{t \in \mathbb{R}} H^{t,p}$ of the partial differential equation $P(x, D)u = f$ on $\mathbb{R}^n$ is in $H^{s+m,p}$.

14.2.   Let $m \geq 0$ and $\sigma \in S^m$ be an elliptic symbol. Let $f \in L^p(\mathbb{R}^n), 1 < p < \infty$, and u be any solution in $\bigcup_{t \in \mathbb{R}} H^{t,p}$ of the pseudo-differential equation $T_\sigma u = f$ on $\mathbb{R}^n$.

(i) Prove that $u \in L^p(\mathbb{R}^n)$.

(ii) Prove that $(u, T_\sigma^* \varphi) = (f, \varphi)$ for all $\varphi \in S$.

(iii) Prove that there exists a sequence $\{\varphi_k\}$ of functions in S such that $\varphi_k \to u$ in $L^p(\mathbb{R}^n)$ and $T_\sigma \varphi_k \to f$ in $L^p(\mathbb{R}^n)$ as $k \to \infty$.

15. WEAK SOLUTIONS OF PSEUDO-DIFFERENTIAL EQUATIONS

We give in this chapter a result on the existence of weak solutions in $L^p(\mathbb{R}^n)$, $1 < p < \infty$, of pseudo-differential equations on $\mathbb{R}^n$. We begin with the definition of a weak solution.

Definition 15.1. Let $\sigma \in S^m$, $m > 0$, and $f \in L^p(\mathbb{R}^n)$, $1 < p < \infty$. A function u in $L^p(\mathbb{R}^n)$ is said to be a *weak solution* of the pseudo-differential equation $T_\sigma u = f$ on $\mathbb{R}^n$ if

$$(u, T_\sigma^* \varphi) = (f, \varphi), \quad \varphi \in \mathcal{S},$$

where T_σ^* is the formal adjoint of T_σ introduced in Chapter 8.

From the definition of the maximal operator $T_{\sigma,1}$ of T_σ given in Chapter 13, it is obvious that the following proposition is true.

Proposition 15.2. Let $\sigma \in S^m$, $m > 0$, and $f \in L^p(\mathbb{R}^n)$, $1 < p < \infty$. Then a function u in $L^p(\mathbb{R}^n)$ is a weak solution of the pseudo-differential equation $T_\sigma u = f$ on $\mathbb{R}^n$ if and only if $u \in \mathcal{D}(T_{\sigma,1})$ and $T_{\sigma,1} u = f$.

For any $\sigma \in S^m$, $m > 0$, the following theorem characterizes the functions f in $L^p(\mathbb{R}^n)$, $1 < p < \infty$, for which the pseudo-differential equation $T_\sigma u = f$ on $\mathbb{R}^n$ has a weak solution u in $L^p(\mathbb{R}^n)$.

Theorem 15.3. *Let $\sigma \in S^m$, $m > 0$, and $f \in L^p(\mathbb{R}^n)$, $1 < p < \infty$. Then the pseudo-differential equation $T_\sigma u = f$ on $\mathbb{R}^n$ has a weak solution u in $L^p(\mathbb{R}^n)$ if and only if there exists a positive constant C such that*

$$|(f, \varphi)| \le C\|T_\sigma^* \varphi\|_{p'}, \quad \varphi \in S, \tag{15.1}$$

where p' is the conjugate index of p.

Proof: Suppose that $T_\sigma u = f$ on $\mathbb{R}^n$ has a weak solution u in $L^p(\mathbb{R}^n)$. Then, by Definition 15.1,

$$(f, \varphi) = (u, T_\sigma^* \varphi), \quad \varphi \in S.$$

Hence, by Hölder's inequality,

$$|(f, \varphi)| \le \|u\|_p \|T_\sigma^* \varphi\|_{p'}, \quad \varphi \in S,$$

and the inequality (15.1) holds with $C = \|u\|_p$. Conversely, suppose that the inequality (15.1) is true. Let W be the subspace of $L^{p'}(\mathbb{R}^n)$ defined by

$$W = \{w \in L^{p'}(\mathbb{R}^n) : T_\sigma^* \varphi = w \text{ for some } \varphi \in S\}.$$

We define the linear functional $F : W \to \mathbb{C}$ by

$$Fw = (\varphi, f), \quad w \in W,$$

where φ is any function in S with the property that $T_\sigma^* \varphi = w$. To see that the defintion of $F : W \to \mathbb{C}$ is independent of the function φ, let φ_1 and φ_2 be functions in S such that $T_\sigma^* \varphi_1 = w$ and $T_\sigma^* \varphi_2 = w$. Then, by (15.1),

$$|(\varphi_1 - \varphi_2, f)| \le C\|T_\sigma^*(\varphi_1 - \varphi_2)\|_{p'} = 0.$$

Hence $(\varphi_1, f) = (\varphi_2, f)$ and this proves that the choice of the function φ is irrelevant to the definition of $F : W \to \mathbb{C}$. Since, by (15.1),

$$|Fw| = |(\varphi, f)| \leq C\|T_\sigma^* \varphi\|_{p'} = C\|w\|_{p'} , \quad w \in W ,$$

it follows that $F : W \to \mathbb{C}$ is a bounded linear functional. Hence, using the Hahn-Banach theorem and the Riesz representation theorem, we can find a function u in $L^p(\mathbb{R}^n)$ such that

$$Fw = (\varphi, f) = (w, u) , \quad w \in W , \tag{15.2}$$

where φ is any function in $\mathcal{S}$ satisfying $T_\sigma^* \varphi = w$. Since $\{T_\sigma^* \varphi \; : \; \varphi \in \mathcal{S}\}$ is obviously a subspace of W, it follows from (15.2) that

$$(\varphi, f) = (T_\sigma^* \varphi, u) , \quad \varphi \in \mathcal{S} ,$$

and hence, by Definition 15.1, u is a weak solution in $L^p(\mathbb{R}^n)$ of the pseudo-differential equation $T_\sigma u = f$ on $\mathbb{R}^n$.

Exercises

15.1. Let $\sigma \in S^m$, $m > 0$, and let u and f be in $L^p(\mathbb{R}^n)$, $1 < p < \infty$. Prove that u is a solution of $T_\sigma u = f$ on $\mathbb{R}^n$ in the distribution sense if and only if u is a weak solution of $T_\sigma u = f$ on $\mathbb{R}^n$.

15.2. Let $\sigma \in S^m$, $m > 0$, be an elliptic symbol, and let $f \in L^p(\mathbb{R}^n)$, $1 < p < \infty$. Prove that every weak solution u in $L^p(\mathbb{R}^n)$ of the pseudo-differential equation $T_\sigma u = f$ on $\mathbb{R}^n$ is in $H^{m,p}$.

15.3. Let $s > 0$ and J_{-s} be the pseudo-differential operator defined in Chapter 11. Let q be any real-valued and nonnegative function on $\mathbb{R}^n$ such that

$$\sup_{x \in \mathbb{R}^n} |(D^\alpha q)(x)| < \infty$$

for all multi-indices α. Prove that the pseudo-differential equation $J_{-s}u + qu = f$ on $\mathbb{R}^n$ has a weak solution u in $L^2(\mathbb{R}^n)$ for every function u in $L^2(\mathbb{R}^n)$.

15.4.   Let $\sigma(x, \xi) \in S^m$, $m > 0$, be such that σ is independent of $x \in \mathbb{R}^n$. Prove that the pseudo-differential equation $T_\sigma u = f$ on $\mathbb{R}^n$ has a unique weak solution u in $L^2(\mathbb{R}^n)$ for all functions f in $L^2(\mathbb{R}^n)$ if and only if there exists a positive constant C such that

$$|\sigma(\xi)| \geq C, \quad \xi \in \mathbb{R}^n .$$

BIBLIOGRAPHY

1. S. Agmon, A. Douglis and L. Nirenberg, *Estimates near the boundary for solutions of elliptic partial differential equations satisfying general boundary conditions,* I, Comm. Pure Appl. Math. 12 (1959), 623–727.

2. Q. Fan and M. W. Wong, *A characterization of Fredholm pseudo-differential operators,* J. London Math. Soc. (2) 55 (1997), 139–145.

3. Q. Fan and M. W. Wong, *A characterization of some non-elliptic pseudo-differential operators as Fredholm operators,* Forum Math. 9 (1997), 17–28.

4. A. Friedman, *Advanced Calculus,* Holt, Rinehart and Winston, 1971.

5. P. Hess and T. Kato, *Perturbation of closed operators and their adjoints,* Comment. Math. Helv. 45 (1970), 524–529.

6. L. Hörmander, *Estimates for translation invariant operators in L^p spaces,* Acta Math. 104 (1960), 93–140.

7. H. Kumano-go, *Pseudo-Differential Operators,* MIT Press, 1981.

8. H. L. Royden, *Real Analysis,* Third Edition, Prentice-Hall, 1988.

9. W. Rudin, *Real and Complex Analysis,* Third Edition, McGraw-Hill, New York, 1987.

10. X. Saint Raymond, *Elementary Introduction to the Theory of Pseudodifferential Operators*, CRC Press, 1991.

11. M. Schechter, *Principles of Functional Analysis*, Academic Press, New York, 1971.

12. M. Schechter, *Modern Methods in Partial Differential Equations*, McGraw-Hill, 1977.

13. M. Schechter, *Spectra of Partial Differential Operators*, Second Edition, North-Holland, Amsterdam, 1986.

14. R. L. Wheeden and A. Zygmund, *Measure and Integral*, Marcel Dekker, 1977.

15. M. W. Wong, *Fredholm pseudo-differential operators on weighted Sobolev spaces*, Ark. Mat. 21 (1983), 271–282.

16. M. W. Wong, *On some spectral properties of elliptic pseudo-differential operators*, Proc. Amer. Math. Soc. 99 (1987), 683–689.

17. M. W. Wong, *Essential spectra of elliptic pseudo-differential operators*, Comm. Partial Differential Equations 13 (1988), 1209–1221.

18. M. W. Wong, *Minimal and maximal operator theory with applications*, Canad. J. Math. 43 (1991), 617–627.

19. M. W. Wong, *Spectral theory of pseudo-differential operators*, Adv. Appl. Math. 15 (1994), 437–451.

INDEX